Advances in Underwater Technology, Ocean Science and Offshore Engineering

Volume 25

Safety in Offshore Drilling

The Role of Shallow Gas Surveys

ADVANCES IN UNDERWATER TECHNOLOGY
OCEAN SCIENCE AND OFFSHORE ENGINEERING

Vol. 1.	Developments in Diving Technology
Vol. 2.	Design and Installation of Subsea Systems
Vol. 3.	Offshore Site Investigation
Vol. 4.	Evaluation, Comparison and Calibration of Oceanographic Instruments
Vol. 5.	Submersible Technology
Vol. 6.	Oceanology
Vol. 7.	Subsea Control and Data Acquisition
Vol. 8.	Exclusive Economic Zones
Vol. 9.	Stationing and Stability of Semi-submersibles
Vol. 10.	Modular Subsea Production Systems
Vol. 11.	Underwater Construction: Development and Potential
Vol. 12.	Modelling the Offshore Environment
Vol. 13.	Economics of Floating Production Systems
Vol. 14.	Submersible Technology: Adapting to Change
Vol. 15.	Technology Common to Aero and Marine Engineering
Vol. 16.	Oceanology '88
Vol. 17.	Energy for Islands
Vol. 18.	Disposal of Radioactive Waste in Subsea Sediments
Vol. 19.	Diverless and Deepwater Technology
Vol. 20.	Subsea International '89: Second Generation Subsea Production Systems
Vol. 21.	NDT: Advances in Underwater Inspection Methods
Vol. 22.	Subsea Control and Data Acquisition: Technology and Experience
Vol. 23.	Subtech '89. Fitness for Purpose
Vol. 24.	Advances in Subsea Pipeline Engineering and Technology
Vol. 25.	Safety in Offshore Drilling. The Role of Shallow Gas Surveys

CONFERENCE PLANNING COMMITTEE

D.A. Ardus, *British Geological Survey*
C.D. Green, *Shell Internationale Petroleum Maatschappij*
J. Pritchard, *Society for Underwater Technology*

Advances in Underwater Technology, Ocean Science and Offshore Engineering

Volume 25

Safety in Offshore Drilling

The Role of Shallow Gas Surveys

Proceedings of an International Conference
(Safety in Offshore Drilling)
organized by the Society for Underwater Technology
and held in London, U.K., April 25 & 26, 1990.

edited by

D.A. Ardus

British Geological Survey

and

C.D. Green

SIPM, The Netherlands

KLUWER ACADEMIC PUBLISHERS

DORDRECHT / BOSTON / LONDON

Library of Congress Cataloging-in-Publication Data

Safety in offshore drilling : the role of shallow gas surveys / edited by D.A. Ardus, C.D. Green.
p. cm. -- (Advances in underwater technology, ocean science, and offshore engineering ; v. 25)
Proceedings of a conference held in London, Jan. 23-24, 1990 and organized by Society for Underwater Technology.
ISBN-13: 978-94-010-6785-0 e-ISBN-13: 978-94-009-0669-3
DOI: 10.1007/978-94-009-0669-3

1. Gas well drilling--Safety measures--Congresses. 2. Gas reservoirs--Congresses. 3. Seismic prospecting--Congresses. I. Ardus, D. A. II. Green, C. D., 1948- . III. Society for Underwater Technology. IV. Series.
TN871.3.S26 1990
622'.3381--dc20 90-5120

ISBN-13: 978-94-010-6785-0

Published by Kluwer Academic Publishers,
P.O. Box 17, 3300 AA Dordrecht, The Netherlands.

Kluwer Academic Publishers incorporates
the publishing programmes of
D. Reidel, Martinus Nijhoff, Dr W. Junk and MTP Press.

Sold and distributed in the U.S.A. and Canada
by Kluwer Academic Publishers,
101 Philip Drive, Norwell, MA 02061, U.S.A.

In all other countries, sold and distributed
by Kluwer Academic Publishers Group,
P.O. Box 322, 3300 AH Dordrecht, The Netherlands.

Printed on acid-free paper

Contents

Society for Underwater Technology

The Society was founded in 1966 to promote the further understanding of the underwater environment. It is a multi-disciplinary body with a worldwide membership of scientists and engineers who are active or have a common interest in underwater technology, ocean science and offshore engineering.

Committees

The Society has a number of Committees to study such topics as:

Diving and Submersibles
Offshore Site Investigation and Geotechnics
Environmental Forces and Physical Oceanography
Ocean Resources
Subsea Engineering and Operations
Education and Training

Conference and Seminars

An extensive programme is organized to cater for the diverse interests and needs of the membership. An annual programme usually comprises four conferences and a much greater number of one-day seminars plus evening meetings and an occasional visit to a place of technical interest. The Society has organized over 100 seminars in London, Aberdeen and other appropriate centres during the past decade. Attendance at these events is available at significantly reduced levels of registration fees for Members or staff of Corporate Members.

Publications

Proceedings of the more recent conferences have been published in this series of *Advances in Underwater Technology, Ocean Science and Offshore Engineering*. These and other publications produced separately by the Society are available through the Society to members at a reduced cost.

Journal

The Society's quarterly journal *Underwater Technology* caters for the whole spectrum of the inter-disciplinary interests and professional involvement of its readership. It includes papers from authoritative international sources on such subjects as:

Diving Technology and Physiology

Civil Engineering
Submersible Design and Operation
Geology and Geophysics
Subsea Systems
Naval Architecture
Marine Biology and Pollution
Oceanography
Petroleum Exploration and Production
Environmental Data

An Editorial Board has responsibility for ensuring that a high standard of quality and presentation of papers reflects a coherent and balanced coverage of the Society's diverse subject interests; through the Editorial Board, a procedure for assessment of papers is conducted.

Endowment fund

A separate fund has been established to provide tangible incentives to students to acquire knowledge and skills in underwater technology or related aspects of ocean science and offshore engineering. Postgraduate students have been sponsored to study to MSc level and subject to the growth of the fund it is hoped to extend this activity.

Awards

An annual President's Award is presented for a major achievement in underwater technology. In addition there is a series of sponsored annual awards by some Corporate Members for the best contribution to diving operations, oceanography, diverless intervention technology and the best technical paper in the Journal.

FURTHER INFORMATION

If you would like to receive further details, please contact
Society for Underwater Technology, The Memorial Building, 76 Mark Lane, London EC3R 7JN.
Telephone: 071-481 0750; Telex: 886841 I Mar E G; Fax: 071-481 4001.

Part I
Introduction

CURRENT DRILLING PRACTICE AND THE OCCURRENCE OF SHALLOW GAS

Paul K Prince

BP Exploration

INTRODUCTION

Shallow gas presents a major worldwide problem to the drilling industry. The records show that approximately one third of all blowouts have been caused by shallow gas. This proportion is equally applicable over the recent past. During the period June 1988 to June 1989 shallow gas was responsible for four of the twelve blowouts that occurred. Fifteen people were killed in three of the four shallow gas blowouts. Drilling rigs were lost in all four incidents.

Planning a well to reduce the chance of a shallow gas blowout depends to a major extent on the provision of an accurate prognosis of the geology and gas risk in shallow sections of the well.

Volume 25: Safety in Offshore Drilling, 3–25.

The purpose of this paper is to illustrate to those involved in preparing the shallow prognosis the nature of the shallow gas problem in drilling operations and the variety of precautionary drilling practices which can be adopted depending upon the estimated risk.

THE NATURE OF THE SHALLOW GAS PROBLEM

Shallow gas kicks cause a unique problem in drilling operations because they cannot be controlled using conventional techniques. Conventional secondary well control techniques for handling a flow of formation fluid into the wellbore require the well to be closed in at the wellhead to stop the flow. The influx of fluid is then circulated out and the well displaced to a drilling mud of sufficient density to prevent further flow from the offending formation. The action of closing in at the wellhead results in additional pressure being applied to the entire wellbore.

In shallow formations with low fracture pressures the action of closing in the well on an influx of gas is highly likely to cause the open hole formations to fracture. This may subsequently result in an underground blowout to a shallow formation or in the formation broaching to the surface (Figure 1). The result of formations broaching and causing a suface crater is catastrophic for bottom supported drilling rigs such as Jack Ups and Production Platforms. Since the well cannot be closed in the gas must be allowed to flow in a safe manner until the producing zone is isolated or depleted.

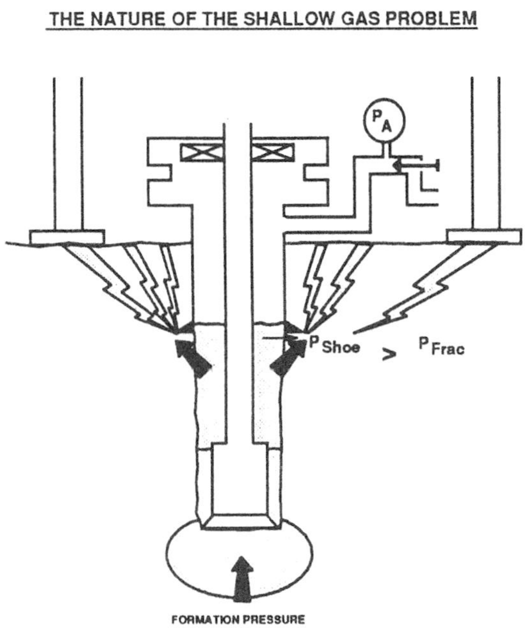

Figure 1.

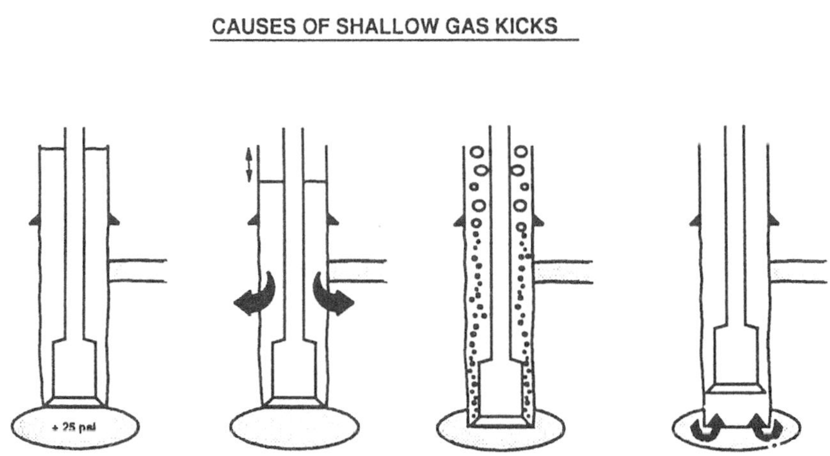

Figure 2.

Further problems can arise when the gas flow is routed through pipework on the drilling installation.

Gas flowing from the producing formation carries with it a high volume of abrasive sand particles and boulders often found in shallow formations. This combination of material will have a tendency to erode or plug any pipework through which it flows. The rate of erosion or plugging is dependent on the design of the pipework. The result of erosion of the flow diversion system could be the release of gas to areas of the drilling installation where there is a potential for explosive ignition of the gas. The effect of a blocked or partially blocked flow path is the same as closing in the well and can have the same castastrophic results.

CAUSES OF SHALLOW GAS KICKS

A shallow gas kick, like any other kick, will occur when the pressure exerted by the drilling mud is less than the pressure of the fluid in the open hole formations. In shallow hole sections the overbalance margin by which the hydrostatic pressure of the drilling mud exceeds the formation pressure is generally less than in deeper formations. In order to generate an overbalance above normal formation pressure of 150 psi at 3000m a mud weight of 1.11sg is required. To generate the same overbalance at 500m would require a mud weight of 1.285sg. However, the low formation fracture gradients in shallow formations limit the maximum mud weight that can be used.

In the shallow sections of many wells a mud weight of 1.285sg would cause the formation to fracture resulting in loss of drilling mud and a consequent reduction in the hydrostatic pressure exerted by the mud column. A more typical shallow section mud weight of 1.15sg would give an overbalance of only 54 psi at 500m, considerably less than in deeper drilling operations.

The relatively small reduction in hydrostatic pressure necessary to allow shallow gas to flow into the wellbore can occur in several ways (Figure 2).

Lost Circulation

If the hydrostatic pressure exerted on the open hole formations by the mud column is greater than the formation fracture pressure then drilling mud will be lost from the wellbore to the formation. The severity of the mud losses may be such that the hole cannot be kept full. If this occurs the height of the mud column and consequently the hydrostatic pressure exerted by the mud column will be reduced.

Gas Cut Mud

When gas bearing formations are penetrated the gas which was present in the formation which has been cut by the drill bit is released into the drilling mud. The drilled gas is circulated out of the well with the drilling mud.

The volume of the drilled gas increases as it's pressure decreases when it is circulated up the annulus. The reduction in mud weight caused by the gas cutting depends on the volume to which the gas has expanded. The reduction is therefore not uniform throughout the mud column. The most significant expansion occurs close to the top of the well and so the measured reduction in mud weight at surface is not representative of the reduction in mud weight throughout the mud column in the wellbore. The total reduction in hydrostatic pressure exerted by the mud column is not then as large as might be indicated by measuring the density of gas cut mud at surface. When drilling deeper formations the reduction in hydrostatic pressure is usually less than the overbalance margin.

However, the reduction in hydrostatic pressure at the top of the mud column may be sufficient to allow gas from shallow open hole sections to flow into the wellbore.

Gas cut mud can also be caused by small volumes of gas which are swabbed into the wellbore when pulling drillpipe to make a connection or to change the bottom hole assembly.

Swabbing

When the drillstring is raised to make a connection or pulled out of the hole it acts to a certain extent like a piston. The effect of the piston like action is to reduce the pressure below the bit and to draw in fluid from the formation.

The reduction in bottom hole pressure caused by swabbing will depend on the annular clearances between the drillstring and the hole, the viscosity and gel strength of the drilling mud and the speed at which the drillstring is pulled out of the hole.

Not Filling the Hole

When drillpipe is pulled out of the hole the level of drilling mud will fall in proportion to the steel volume of the pipe that has been removed. If additional mud is not pumped into the well then the height of the mud column decreases and as a result the hydrostatic pressure exerted by the drilling mud will be reduced.

Insufficient Mud Weight

If the formation fluid is at a pressure higher than the hydrostatic pressure exerted by a full column of uncontaminated drilling mud then mud will flow into the wellbore. This situation would occur when the actual formation pressure is higher than anticipated when planning the well.

WELL PLANNING

The first step in avoiding a shallow gas kick is a high quality seismic survey.

The interpreted results of the survey should provide the drilling engineer with a prediction of the shallow geology and associated shallow gas risk of the proposed well location. This information can be used in both the design of the well and the planning of drilling procedures for the shallow sections of the well. If the risk of drilling through shallow gas formation is high it may be possible to move the surface location of the well to a lower risk site. If the target formations for the well do not allow a vertical well to be drilled from the new location a more costly directional well would then be required.

Dependent on the depth of the shallow gas formation and the local formation fracture gradient it may be possible to design the well to reduce the risk of a shallow gas kick occurring. Casing off the formations above the shallow gas zone before drilling into it might have two benefits. First, it may allow conventional secondary well control techniques to be used in the event of an influx from the gas bearing formation ie to close in at the wellhead and displace the well to a heavier drilling mud without fracturing the open hole formations.

Second, it may allow drilling mud of a higher density to be used when drilling the gas bearing formation. This would exert a higher overbalance pressure on the gas bearing formation so reducing the risk of a kick being caused by any of the previously described phenomena.

DRILLING PROCEDURES

The risk of causing a shallow gas kick by many of the previously described phenomena can be reduced by adopting suitable precautionary drilling procedures.

Rate of Penetration

The rate at which new formation is cut by the drill bit will affect both the amount of drilled cuttings and drilled gas in the wellbore. If the mud column is loaded with cuttings the effective mud weight is higher than that of a clean mud and so the hydrostatic pressure exerted by the mud column is increased. This increase may be sufficient to induce lost circulation. The rate at which drilled gas enters the drilling mud will affect the degreee of gas cutting of the mud and the distribution of gas in the mud column. In order to reduce the risk of a shallow gas kick being caused by drilled gas or lost circulation the instantaneous rate of penetration should be controlled below a predetermined maximum. A commonly used instantaneous maximum is 30 metres per hour.

Mud Circulation Rate

The rate at which mud is circulated up the annulus will also affect the concentration of drilled cuttings and gas cut mud in the wellbore.

The higher the circulating rate the faster cuttings and gas are removed from the well and so the lower the risk of causing a kick. However the maximum rate at which drilling mud can be circulated is not limited only by the mud pump power available. High circulating rates through the annulus have two other effects. They tend to erode the open hole causing undesirable washed out hole. They also increase the annular friction losses which act on the bottom of the well and so increase the effective weight of the drilling mud. However the annular friction loss is also dependent on the length of the wellbore and the size of the anullii between the drillstring and the wellbore. In shallow sections the length of the annulus is short and the size of the annulus is usually large so that annular friction losses are negligible. This may not be the case when a pilot hole is drilled.

Pilot Hole

The size of the hole that is drilled through the formation will affect the concentration of cuttings and drilled gas in the wellbore. The final size of hole required depends upon the size of casing to be run. In shallow sections of the well the hole size required is usually between 24 and 36 inches. By drilling a smaller pilot hole and then opening it up to the required size in a second or third pass the concentration of cuttings and gas in the hole can be reduced. The smaller annular capacity of the pilot also produces higher annular velocities on the first pass. A possible disadvantage in drilling a small pilot hole is that the smaller annular capacity may lead to a greater swabbing effect.

On exploration wells some operators also drill a pilot hole to allow electric logs to be run across the shallow section of the well. These logs will provide a definitive indication of the presence of shallow gas and give valuable feedback to the shallow seismic interpreter.

Tripping Procedures

Continuously filling the well as drillpipe is pulled out of the hole is standard drilling practice. This is usually achieved by circulating mud across the well from a small volume trip tank as pipe is removed.

In order to reduce the swabbing effect when pulling drillpipe the mud is conditioned to reduce the viscosity and gel strength to the lowest acceptable values prior to starting the trip out. The speed at which the pipe is pulled is also restricted. To further reduce the swabbing effect some operators circulate through the drillstring as it is being removed from the well.

Circulating Out Gas Cut Mud

If an unusually high degree of gas cutting of the mud is observed at surface drilling should be stopped. Circulation of the drilling mud is continued through rig mounted de-gassing equipment until the gas is removed from the well. When drilling is resumed it may be necessary to reduce the rate of penetration through the gas bearing formation.

Floating Drilling Rigs

On floating drilling rigs it is possible to further reduce the risk of shallow gas kicks occurring as a result of several of the previously described causes, namely lost circulation, not filling the hole and gas cut mud. This can be achieved by drilling the shallow section of the well, where conventional secondary well control techniques cannot be used, without a marine riser. The marine riser is the conduit between the wellhead on the seabed and the rig. When drilling without a riser the mud from the annulus travels only as far as the seabed. There are several advantages in using this technique. The length of the mud column is reduced and so the risk of losses induced by loading of the annulus with cuttings is reduced. If lost circulation does occur the presence of the seawater will ensure that the hole is kept full. When pulling out of the hole there is no possibility of the hole not remaining full. The effect of gas cut mud will be reduced as it leaves the wellbore before it is fully expanded.

When drilling riserless the mud used is normally seawater with occasional slugs of viscous mud to ensure that the hole is cleaned of cuttings. If the shallow geological prognosis indicates that a more sophisticated or weighted mud system is required for wellbore stability or because of abnormally pressured formations then a riser must be deployed.

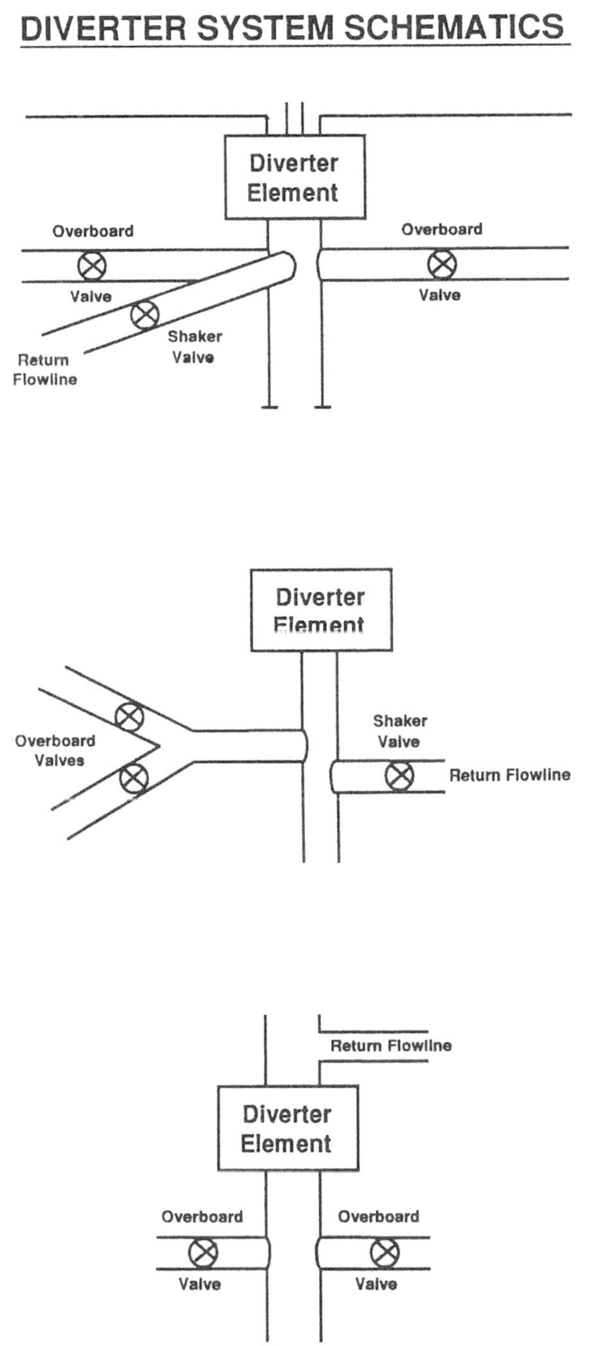
DIVERTER SYSTEM SCHEMATICS
Diverter Element
Overboard
Valve
Overboard
Valve
Shaker Valve
Return Flowline
Diverter Element
Overboard Valves
Shaker Valve
Return Flowline
Return Flowline
Diverter Element
Overboard
Valve
Overboard
Valve

Figure 3.

SHALLOW GAS KICK CONTROL EQUIPMENT AND PROCEDURES

Shallow gas kicks will occur in situations when suitable precautions are not taken or when the precautions are not sufficient to stop the kick. Equipment and procedures must therefore be in place to allow gas to flow from the well in a safe manner until the producing zone is isolated or depleted or at least until the risk to personnel can be minimised.

Diverter System

The diverter system on a drilling rig is the package of equipment necessary to safely divert a flow of gas away from the well to a non hazardous area. The system usually comprises of the following parts:-
(Figure 3)

1. One or two pipelines (diverter lines) running from the wellhead to a non hazardous area.
2. An elastomeric sealing element to close off the top of the well (diverter element).
3. A number of valves as required to divert returning fluid from the normal flow pipeline to the diverter line.
4. A hydraulic control system to operate the diverter element and the valves.

When the diverter is activated the constituent parts should operate in the following sequence.

1. The valve on the selected diverter line opens
2. The valve on the return flowline closes
3. The diverter element closes around the drillpipe

Following this sequence ensures that the well is never closed in. There are several different configurations of diverter system in use. However the principle of the closing sequence should always be that the flow from the well is diverted into the diverter line and all other lines from the well are shut off. This should be achieved without closing in the well.

There are specific design requirements for each constituent part of the diverter system.

Diverter Lines

The diverter lines must allow gas and the sand and other debris carried with it to flow away from the well while causing the minimum back pressure on the well. The back pressure generated in the diverter system acts on the entire wellbore. A significant amount of back pressure will cause the open hole formations to fracture in the same way as closing in the well. Back pressure in the diverter line will be caused by the friction force exerted against the fluid flow and by plugging or partial plugging of the line by debris.

The friction losses and tendency to plug can both be reduced by making the diverter lines large diameter and as straight as possible. 12 inches is a frequently quoted minimum diameter.

The mixture of gas and sand travelling at high velocity through the lines is highly abrasive. If holes were to be eroded in the diverter line it would result in the release of gas to hazardous areas of the rig where there is a potential for explosive ignition. The greater the wall thickness of the diverter line the more resistant to erosion it will be.

Straight lines will be more erosion resistant than those with bends.

Diverter Element

The diverter element should be able to close rapidly and seal on the various diameters of the drillstring which may be across the wellhead. It must be able to contain any wellhead pressure caused by the gas flow and be resistant to the abrasive nature of the flow.

Diverter Line Valves

The diverter line valves when opened should not restrict the flow or be liable to cause the line to plug. They should therefore be fast operating full opening valves.

Hydraulic Control System

The hydraulic control system used to activate the diverter system should interlock each part of the system so that the correct operating sequence must be followed and that the well cannot be closed in. Ideally the entire sequence should be activated by throwing a single lever so reducing the possibility of human error in executing the divert procedure.

Diverter System Reliability

A survey by A.S. Veritas Research (3) in 1986 suggested that the failure rate of diverter equipment when used in shallow gas blowout situations was of the order of 50%.

The main causes of failures were erosion and improper functioning of valves.

The long term objective in diverter system design must be to allow the well to flow safely until the shallow gas zone has depleted or primary control can be re-established. However where the equipment installed cannot meet this objective the minimum requirement must be to allow the well to flow safely until personnel can be evacuated to a safe area.

When drilling from a floating rig with a riser and rig mounted diverter system there is a further potential weak point in the system.

The telescopic joint in the riser which allows the rig to heave up and down depends on one or two elastomeric seals for pressure integrity. If these seal are worn gas will leak at an increasing rate as the seals wash out.

A floating rig does have the advantage of being able to move away from the well in the event of a failure of the diverter system. In fact some operators will make preparation to move off location immediately the well has been diverted.

Shallow Gas Kicks When Drilling Riserless

Riserless drilling was previously discussed as a method of reducing the risk of shallow gas kicks occurring as a result of several causes. In the event of a kick occurring riserless drilling has a further major advantage over rig mounted diverter systems in that there is no conduit through which the gas can flow directly to the rig. There are therefore no associated diverter system problems. The plume formed as gas escapes from the well at the seabed will most likely be directed away from the rig by the prevailing current.

In the event that gas is released at sea level below the rig any wind will tend to disperse the gas before it reaches a semisubmersible rig. If gas does start to build up the rig can be moved off location. This operation will be simplified without the necessity to disconnect a riser.

The rate at which gas flows from the well in a fully developed blowout may be reduced in the riserless case by the ever present hydrostatic pressure exerted by the seawater. A blowout through a riser may unload all liquid from above the producing formation.

Concern used to be expressed that the gas plume would cause a loss of buoyancy below a floating vessel causing it to sink. Research work (4) and past experience have shown that this is not the case. There have been cases where drillships have sunk during shallow gas kicks. These incidents were caused not by loss of buoyancy as a result of the gas plume but by water flowing over the side of the ship and entering the internal voids. This situation can be avoided by keeping all hatches closed while drilling shallow sections of the well (Figure 4).

NEW DEVELOPMENTS

The drilling industry continues to invest in research and development to improve our ability to control shallow gas kicks. Some of the new developments arising from this investment are designed to improve our understanding of the nature of shallow gas blowouts while others result in new hardware. All have a potential effect on the procedures adopted.

SHALLOW GAS KICK WHEN DRILLING RISERLESS

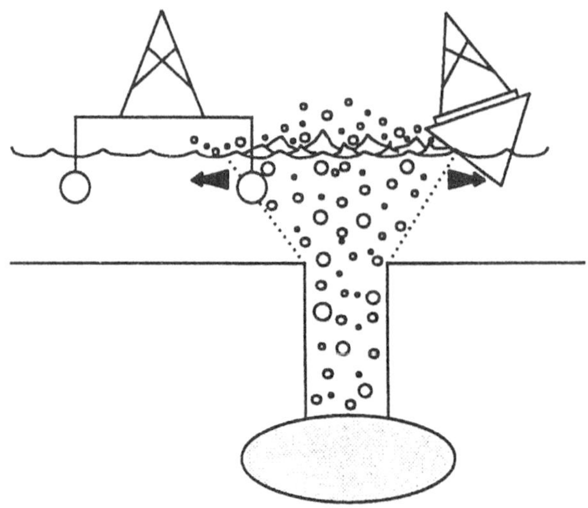

Figure 4.

SUBSEA DIVERTER

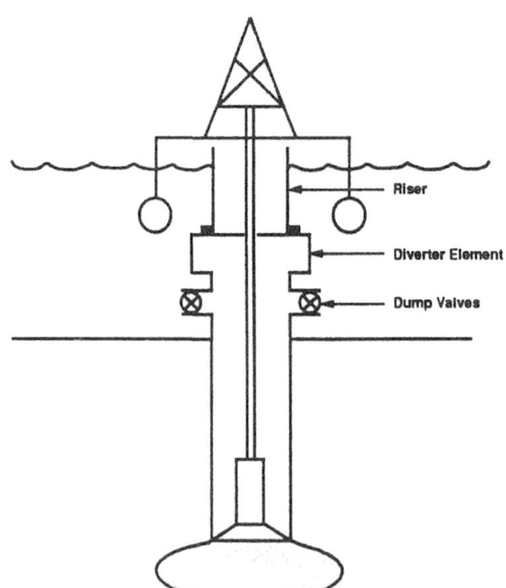

Figure 5.

Subsea Diverter

When drilling shallow sections of the well which dictate that a weighted mud system is used it is not practicable to drill riserless from a floating rig. Subsea diverter systems have been developed to divert the gas flow at seabed in this situation. The diverter element is positioned between the wellhead and the riser. An outlet spool with valves is positioned below the diverter element. In the event of a shallow gas kick the diverter element is closed and the outlet valve opened thus diverting the gas flow to the seabed (Figure 5).

Valveless Diverter System

Diverter systems have been developed which remove the requirement for diverter line valves thus reducing the complexity of the control system. This has been achieved in two ways. One, by replacing the diverter line valve with a low pressure rupture disc (2). In these systems the well is closed for a very short period of time until the wellhead pressure bursts the rupture disc and opens the vent line. Two, by using the same mechanism which closes the diverter element to open the vent line.

Downhole Packer

A prototype downhole packer has been developed which is run in the bottom hole assembly above the bit. In the event of a shallow gas kick the packer is inflated thus sealing off the annulus just above the bit. A circulation port can then be opened above the packer to displace the annulus to the appropriate fluid.

Measurement While Drilling Tools

MWD tools able to detect an influx of formation fluid as a change in the properties of the wellbore fluid are currently under development. This form of detection will provide the driller with an additional surface indication that a kick may have occurred and so assist early detection when drilling at all depths. Shallow gas kicks however have historically occurred very rapidly with a short time lag between any surface warning sign and gas reaching surface. The advantages of these tools may not therefore be as great in shallow drilling operations.

REFERENCES

(1) Goins W.C., Ables G.L., (1987). The Causes of Shallow Gas Kicks.
SPE/IADC 16128

(2) Crawley F.K., Thorogood, J.L., (1987). Single vent-line design selected for diverter system. Oil and Gas Journal, Sept 14, 1987.

(3) Nokleberg, L., Schuller R.B., Sontvedt T., (1986). Shallow Gas Kicks, Safety aspects related to diverter systems. A.S. Veritas Research

(4) Moros A., Dand I., (1990). Two phase flow as a result of a subsea blowout and their effect on the stability of structures. OTC 6478

SHALLOW GAS AND GAS SEEPAGES: A DYNAMIC PROCESS?

A.G. Judd

School of Environmental Technology

Sunderland Polytechnic

INTRODUCTION

Because of the implications of shallow gas accumulations to drilling and the integrity of platform foundations, it is important to determine whether gas accumulations can be considered as effectively static, unchanging for the duration of offshore operations, or dynamic. Shallow gas is frequently seen in association with seabed features, such as mud lumps, gas domes and pockmarks. These are considered to have been caused by the escape of gas through the seabed, consequently it can be implied that gas has migrated to, and through, the seabed. This indicates that shallow gas is not static, but it does not suggest a time-scale over which migration occurs. Is shallow gas mobile on a geological or a human time-scale?

During 1977 and 1978 the British Geological Survey (then the Institute of Geological Sciences) undertook repeat analogue

Volume 25: Safety in Offshore Drilling, 27–50.

geophysical surveys of their South Fladen Pockmark Study Area as part of a programme of investigations into pockmarks, the mechanism by which they were formed and their hazard potential (McQuillin *et al*, 1979). During these surveys three anomalously large pockmarks were discovered. One of them, christened the 'Witch's Hole' during a previous survey, is close to an area of acoustic turbidity within the topmost sediments, the Witch Ground Formation.

This paper describes an analysis of data from these and other surveys collected from around the Witch's Hole, and considers whether or not there is any evidence of shallow gas migration, and whether any deductions can be made about the rate of migration.

MIGRATION THROUGH THE SEABED

Evidence of gas migration is mainly confined to seabed or near-seabed features and processes.

Gas plumes

Gas rising from the seabed has been observed as plumes on echo sounder records, although there is a danger of confusion with the similar signature given by shoals of fish. In most instances these occurrences are observed only once, however the example shown in Figure 1 has been observed five times over a six year (1983 to 1989) time period.

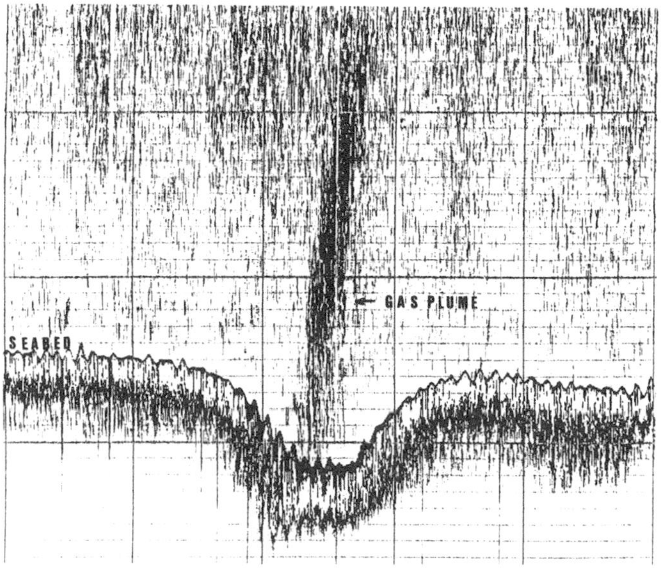

Figure 1. Gas seepage from an active pockmark.

Echo sounder record from an active pockmark in U.K. block 15/25.
Fix marks are approximately 150 m. apart.
Depth interval shown: 105 to 175 m.

Geochemical surveys

Geochemical surveys used in petroleum prospecting take the form of on-line analyses of the hydrocarbon content of near-seabed waters ('Sniffer' surveys) and analyses of the hydrocarbons in the pore spaces and adsorbed on clay particles of seabed sediments collected by coring (Sweeney, 1988). These surveys indicate a positive flux of hydrocarbons, principally methane, and may provide evidence of the source. Faber and Stahl (1984) related the hydrocarbon content of seabed sediments to source rock potential.

Seabed Topographic Features

Fluids rising to and through the seabed are responsible for the formation of several seabed features including mud volcanoes, mud mounds and pockmarks. Of these the most widespread are pockmarks which occur over an area of approximately 100,000 km^2. in the North Sea even though they are restricted to areas in which the seabed sediment is suitable for their formation. Although this implies a vast number of gas seepage sites it is probable that only a small proportion of the pockmarks are currently 'active'.

Methane-derived Carbonates

Carbonate precipitates have been found in association with active methane seepages at several locations in the North Sea and elsewhere (Hovland and Judd, 1988). These are rock-like slabs and blocks exposed on the seabed in pockmarks and seepage areas. They are formed by the cementation of the normal seabed sediments by a precipitate of aragonite and calcite. Carbon isotope studies have shown that their carbon is derived from methane.

Although it is not evidence of current seepage activity, the existence of considerable thicknesses (<25 cm.) of slab-like carbonates in some pockmarks suggests that there has been a flux of methane for a considerable time period.

Bacterial Mats

The presence of bacteria at seepage sites, reported from several locations including some in the North Sea (Hovland and Judd, 1988), is also evidence of a positive flux of gas through the seabed.

THE WITCH'S HOLE

The Witch's Hole is a pockmark measuring about 120 metres across north to south, 90 metres east to west, and 3 metres in depth. It is located in a shallow gas area within the Witch Ground Basin of the U.K. sector of the North Sea, approximately 35 kilometres north east of the Forties field (see Figure 2). It is 'fresh' in appearance in that both within and around the main pockmark feature there are many smaller 'unit' pockmarks (see Figure 3). These are generally up to 2 or 3 metres in diameter, but some are as much as 10 metres across. According to Hovland and Judd (1988) this is indicative of a pockmark which is, or has recently been seeping gas. Indeed the feature was described as 'active' during two separate surveys in 1975.

The following surveys have been conducted in the vicinity of the Witch's Hole:

1) 1975 site survey for well 21/3-2 (approximately 2 kilometres NNE): analogue and digital seismics and side scan sonar; undertaken by Aquatronics for Total.

2) 1975 'Sniffer' survey; undertaken by Interocean for B.P./B.G.S..

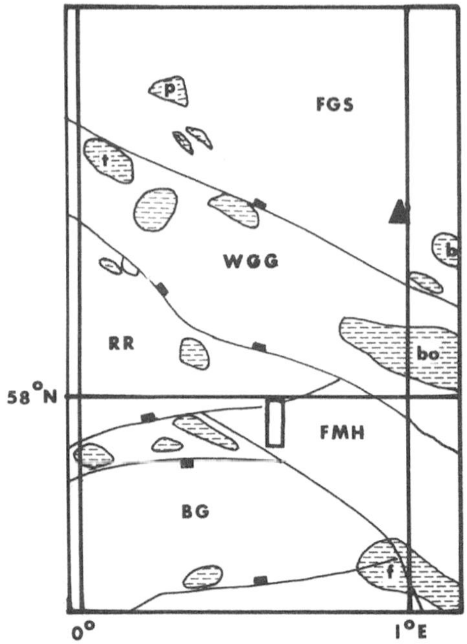

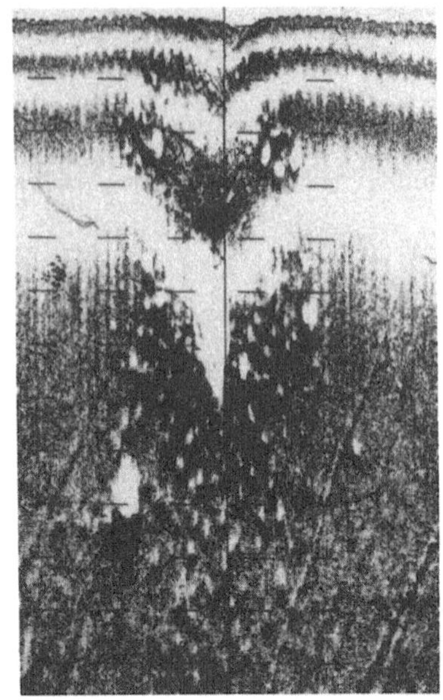

Figure 2.
Location of the Witch's Hole.

▯ Witch's Hole Study Area.

▲ Active pockmark in U.K. block 15/25.

Figure 3. The Witch's Hole

A side scan sonar image from the 1977 survey.

STRUCTURAL FEATURES		OIL FIELDS	
BG	Buchan Graben	b	Balmoral
FGS	Fladen Ground Spur	bo	Bosun
FMH	Forties-Montrose High	f	Forties
RR	Renee Ridge	p	Piper
WGG	Witch Ground Graben	t	Tartan

(details after GECO, 1989)

3) 1977 analogue seismic and side scan sonar; undertaken by Sonarmarine for B.G.S./B.P.

4) 1978 analogue seismic and side scan sonar; undertaken by Sonarmarine for B.G.S.

5) 1987 side scan sonar (two lines); undertaken by Total.

6) 1987 digital seismic (two lines); undertaken by Wimpol for Total.

Unfortunately the overlap between these surveys, both in terms of geographical coverage and equipment compatibility is not ideal, however tentative comparisons can be made.

Seismic Evidence of Shallow Gas

Indications of shallow gas are present in the form of bright spots on digital seismic records and acoustic turbidity on analogue seismic profiles at numerous levels beneath the seabed down to the limits of penetration of the various systems used. These indications are summarised as follows:

1) extensive areas of acoustic turbidity within the very soft silty clays of the Witch Ground Formation from approximately 15 metres below seabed; seen on the analogue seismic surveys of 1975, 1977 and 1978.

2) frequent bright spots within irregularly-bedded reworked glacial silts and clay sands between 55 and 130 metres; seen on the 1975 and 1987 digital seismic surveys.

3) an isolated bright spot within flat-bedded silt and clays at 240 metres; seen on only one of the 1975 digital seismic lines.

4) an extensive area of bright spots associated with an erosion surface at approximately 320 to 380 metres; seen mainly on the 1975 digital seismic lines, but also from the 1987 survey.

5) bright spots at a depth of approximately 670 metres; seen only on digital seismic lines from the 1987 survey (below the effective depth of penetration of the 1975 survey).

The conclusion that can be drawn from this coincidence of accumulations at so many levels is that the gas is rising from a source at greater depth.

In this part of the northern North Sea the most likely source rock is the Kimmeridge Clay. The Witch's Hole lies over the Montrose-Forties High (Figure 2) where the base-Cretaceous lies at a depth of approximately 3,000 metres, and the Kimmeridge Clay lies even deeper. The onset of maturity occurred early in the Tertiary period. Continuous generation of petroleum since then has evidently resulted in the gradual vertical migration of hydrocarbons, particularly the methane, which has accumulated in successive porous horizons within a principally argillaceous sedimentary sequence.

Repeat Seismic Surveys

It is possible that the migration of gas from one accumulation to another occurs only periodically. In this case the accumulations may vary in extent over time as they are filled from below and empty upwards. It was hoped that such size variations might be mapped by comparing the extent of individual bright spots and areas of acoustic turbidity on seismic records from different surveys. The identification of bright spots and acoustic turbidity is quite subjective, consequently the interpretation that is plotted on a map may be questioned by another person. When this subjectivity is compounded by differences in equipment and by position fixing errors it is inevitable that the comparisons must be tentative.

Digital Surveys

The survey undertaken in 1987 was designed not to replicate the coverage acquired in 1975, but to investigate the possibility that gas lay directly beneath the Witch's Hole; there is only a limited overlap between the survey lines. The equipment specifications were different (3 kilojoule sparker in 1975, 13.5 kilojoule sparker in 1987), the position fixing systems were different (Hi-Fix in 1975, Syledis in 1987) and the processing parameters varied; these factors must be borne in mind when considering the following.

Bright spots have been identified at four levels (see above), but of these only the third (320 to 380 metres; 0.575 seconds) is sufficiently extensive to be comparable. According to the 1975 survey (interpretation by Aquatronics, 1975) these bright spots cover an apparently uninterrupted area of approximately 2.2 square kilometres, only part of which is shown in Figure 4. One of the 1987 survey lines extends into this area, and the bright spots correlate well. There appears to be disagreement between the 1987 east-west line and the ends of the north-south 1975 lines. This may or may not be genuine. One of the points of overlap can be seen on Figures 5a and 5b.

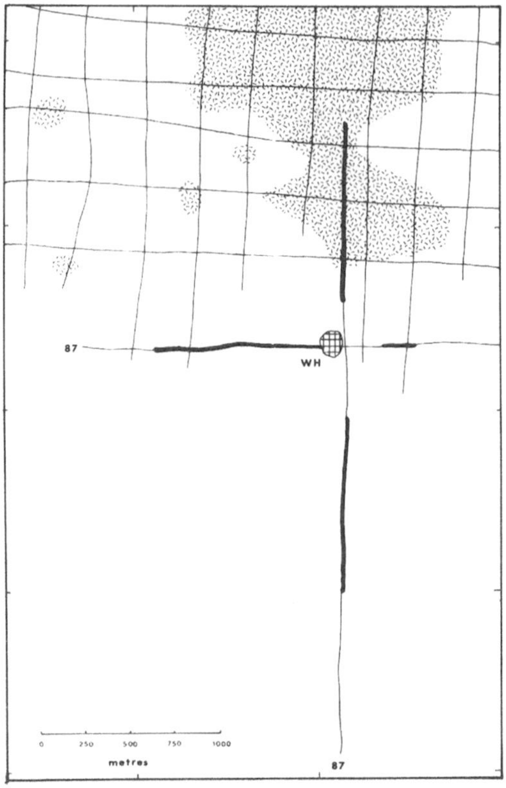

Figure 4. Bright spots around the Witch's Hole 1975 -1987.
Bright spots indicating the presence of shallow gas at a depth of 320 to 380 metres; 0.575 seconds.

WH	Witch's Hole
Shading	extent of bright spots in 1975 (details from Aquatronics, 1975)
———	1975 survey lines
87 ———	1987 survey lines
———	1987 survey lines with bright spots

Analogue Surveys

The 1978 survey consisted of north-south lines with a spacing of 150 metres. In addition nine east-west lines were run close to, and two diagonal lines were run through the Witch's Hole. This comprehensive survey is used as the standard against which the

Figure 5 Digital seismic sections showing bright spots.
5a. 1975 survey (top) 5b 1987 survey (bottom)
The survey lines were run at right angles.
The crossing point is indicated by the vertical line.
Two way travel times of 2 to 7 secs. shown on both sections

other less detailed (1975 and 1977) analogue surveys are compared. It is not considered necessary to show the 1978 survey lines on Figures 6 and 8.

The main area of acoustic turbidity lies to the south and west of the Witch's Hole, the edge being close to the pockmark itself. The acoustic turbidity occurs about 15 metres below the seabed within the topmost sediment layer, the Witch Ground Formation, which is generally about 20-30 metres thick in this area. The acoustic turbidity, which extends beyond the area surveyed, is almost unbroken over an area of at least 4.5 km^2, although there are a few 'windows' and some outlying patches. Windows and patches are generally seen on more than one survey line. Once appropriate towfish layback corrections have been applied there is excellent agreement between different survey lines, so there is a high level of confidence in the distributions shown on Figures 6 and 8. The only area where this confidence is diminished is along the western limit of the acoustic turbidity, roughly parallel to the survey lines, where the interpretation is most subjective.

1977 - 1978 comparison.

The 1977 and 1978 surveys were both undertaken by Sonarmarine who used identical Huntec deep-towed boomer systems, consequently the profiles are directly comparable. However the position fixing systems used were different (Hi-Fix in 1977, Pulse-8 in 1978). In order to ascertain the significance of this the co-ordinates of the Witch's Hole was determined from each of the boomer profiles that crossed it. Good agreement was found between different profiles from individual surveys, and there was an offset of 82 metres between the location as determined from the 1977 and 1978 surveys. Although significant, this is not a major source of error and it will not unduly affect the conclusions of this comparison.

In 1977 the coverage of the area consisted of five north-south lines and a series of short lines radiating from the Witch's Hole. There is close agreement between the surveys with respect to the main area of acoustic turbidity, however there are minor discrepancies in the location of the edges of this area (see Figure 6). Tempting as it is to refer to these as evidence of migration, the systematic errors are such that this cannot be justified. The only places where any reliability might be placed on the interpretation is where profiles from the two surveys cover the same ground and reference can be made to a seabed feature that can be identified on both. The only instance where this happens is on matching lines which pass across the Witch's Hole (see Figures 7a and 7b). There are no grounds for claiming any migration on these profiles.

1975 - 1978 comparison.

The 1975 Aquatronics survey used a hull-mounted 3.5 / 7kHz microprofiler and a high resolution multi-tip sparker. The survey lines were the same as those for the digital survey, so there is very limited coverage of the main area of acoustic turbidity (see Figure 8), however all the survey lines that crossed the area defined in 1978 encountered acoustic turbidity. The agreement is generally good, except for one line near the

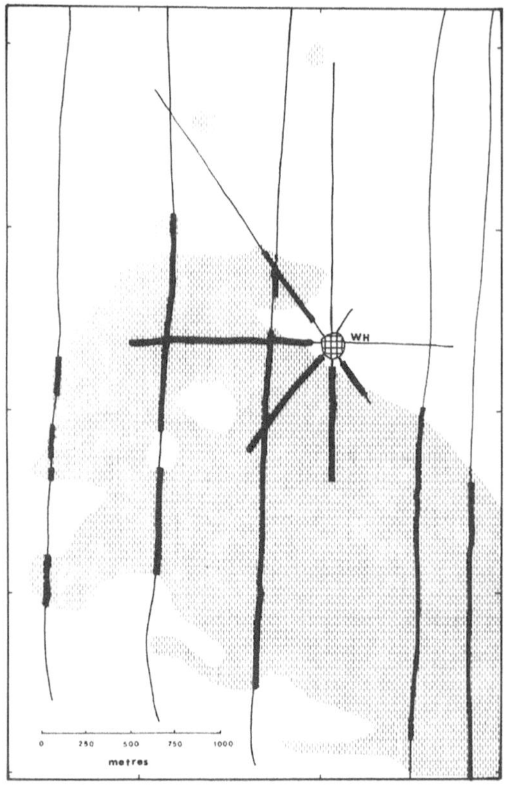

Figure 6.
Acoustic turbidity in the Witch Ground Formation 1977 - 1978.

WH	Witch's Hole
Shading	extent of acoustic turbidity in 1978
——— (thin line)	1977 survey lines
▬▬▬ (thick line)	1977 survey lines with acoustic turbidity

Witch's Hole where acoustic turbidity was not seen in 1978 and one, to the west of the area, where less was seen than expected. These discrepancies may be attributable to either: position fixing errors, the different equipment specifications, interpretative errors, or a genuine change in the position of the acoustic turbidity. However as the acoustic turbidity has been shown to correlate on four lines it is difficult to invoke

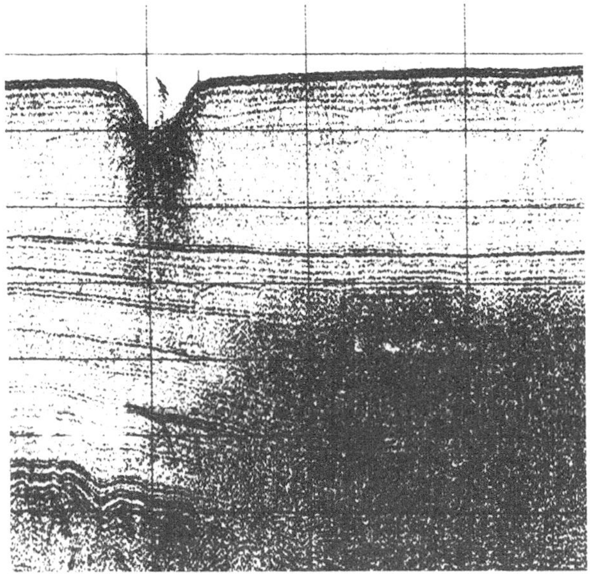

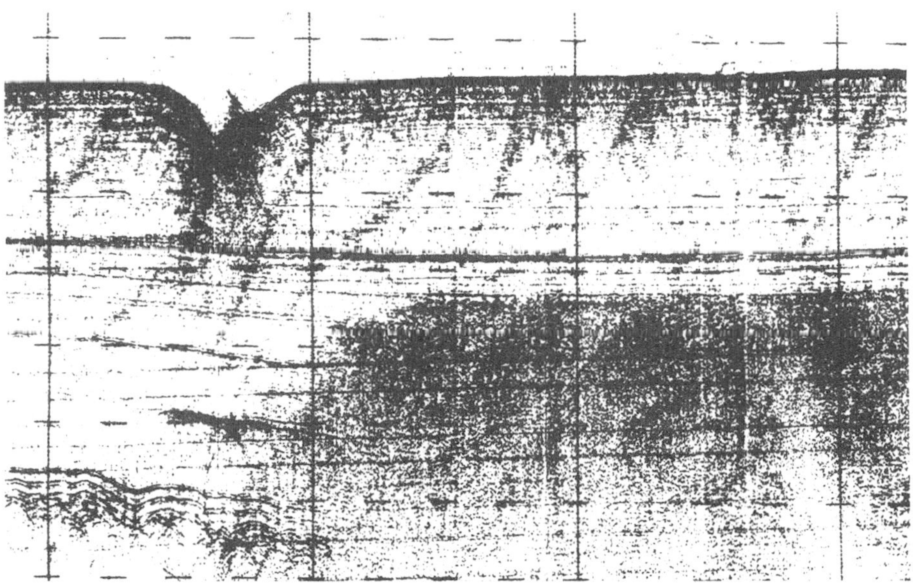

Figure 7. Boomer profiles across the Witch's Hole
7a 1977 survey (top) 7b 1978 survey (bottom)

Profiles run north to south. Acoustic turbidity appears unchanged between 1977 and 1978. Note also the enhanced reflection directly beneath the Witch's Hole and the 'plume' in the Witch's Hole.

N.B. The horizontal scales are different.

any kind of systematic error. Consequently it is concluded that there has been a change in the extent of the area between 1975 and 1978.

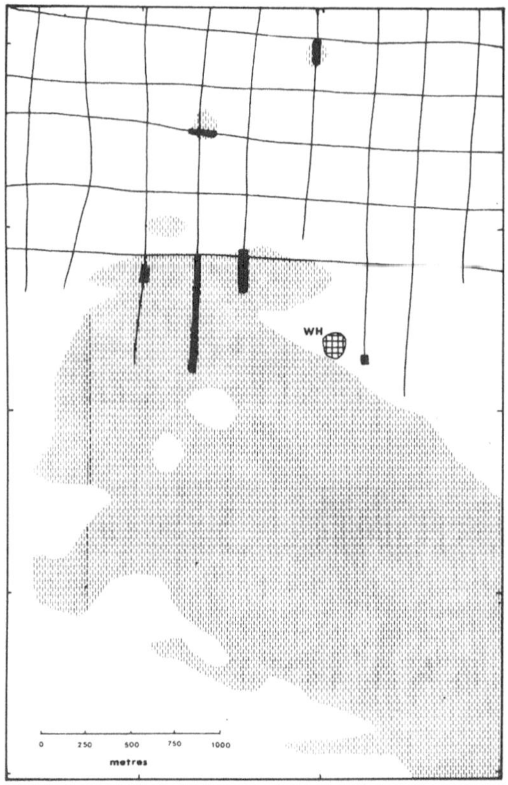

Figure 8
Acoustic turbidity in the Witch Ground Formation 1975 - 1978.

WH Witch's Hole
Shading extent of acoustic turbidity in 1978
——— 1975 survey lines
━━━ 1975 survey lines with acoustic turbidity

Migration Through the Seabed

Over the area of acoustic turbidity there are many pockmarks, however none of them is as large as the Witch's Hole and none has the same 'fresh' appearance. The presence of the gas around, but not directly beneath, suggests that gas may be migrating laterally to the pockmark and then escaping vertically through an unseen 'chimney' to the seabed.

The Witch Ground Formation sediments of this area are very soft to soft muds with subordinate sandy horizons (Stoker *et al*, 1985). They are acoustically transparent on the boomer records, but contain many internal reflections which are thought to represent the sandy layers. On most profiles the top of the acoustic turbidity is flat and coincides with one of these reflections. In many cases, where these reflections slope upwards away from the acoustic turbidity they are enhanced (darker than normal in appearance, i.e. of higher amplitude). It is thought that gas dispersed within the muds, represented by the acoustic turbidity, is rising vertically through the lower part of the formation, its progress being halted at successive sandy layers, through which it migrates laterally towards a chimney. On Figure 7 (a and b) lateral migration towards the Witch's Hole within the sandy layers is indicated by reflection enhancement.

Carbonates and bacterial mats.

As there have been no visual surveys of the Witch's Hole it is not known whether or not bacterial mats are present. Similarly there is no direct evidence of methane-derived carbonates, however the floor of the pockmark has a highly reflective appearance on the side scan sonar records acquired in 1987.

This may be indicative of the presence of carbonate.

Geochemical evidence.

The 'Sniffer' survey undertaken in 1975 (reported by Hovland and Judd, 1988) found anomalous concentrations of methane and total hydrocarbon gases in the bottom waters around the Witch's Hole. The presence of these anomalies suggests that gas is seeping from the sediments into the water column

Sediment Plumes

McQuillin and Fannin (1979) reported that during the 1978 survey plumes of sediment were seen in the water column at heights of up to 10 metres above the seabed on both analogue seismic (boomer) and side scan sonar records. The plumes were seen whilst shooting a series of closely-spaced survey lines near the Witch's Hole. Two separate events were recorded and the plumes dispersed, with the sediment apparently settling to the seabed over a period of a few hours. Figure 9 is another example of a sediment plume recorded during the 1977 survey of the Witch's Hole area.

It is impossible to estimate the volume of sediment within the plumes, or the volume of gas that would have been required to lift it from the seabed, however this evidence suggests that there was gas escape activity during these two surveys. It also supports the contention, expressed by Hovland and Judd (1988) as a result of the detailed mapping of pockmarks from part of the South Fladen area, that pockmark growth occurred during the period between the 1977 and 1978 surveys.

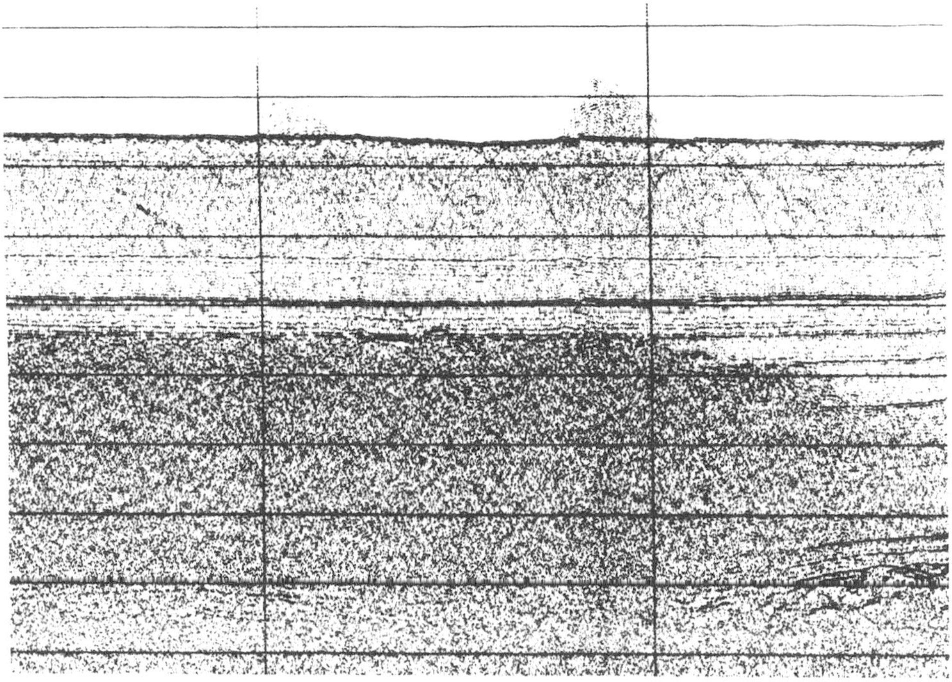

Figure 9. Sediment plume in the water.

Boomer profile from the 1977 survey.
Timing lines approximately 7m. apart.

Gas in the water column?

Both the 1975 surveys described the Witch's Hole as active because of 'plumes' rising from the feature into the water column. A plume can also be seen on both the 1977 and 1978 boomer records (Figures 7a and b). during the side scan sonar investigation of 1987 the towfish collided with this 'plume' and when retrieved it was entangled with trawl netting.

Detailed inspection of the records from this survey showed a sonar target protruding 1.5 to 2.0 metres above the general level of the seabed, indicating a total height of about 5 metres (Figure 10). The target is 5 - 8 metres wide and 22 - 28 metres long. On one of the survey lines two parallel targets are seen lying about two metres apart. One is slightly broader than the other, but does not stand so high above the seabed. The target is centrally located within the pockmark. It presents a 'solid' image at one end, but has regularly-spaced parallel stripes across its width. The target appears to be a wreck, the solid end being either the bow or the stern and the stripes representing the hull cross-members. The hull is lying on its side, the twin targets being the two sides of the ship. It is orientated roughly north-south, parallel with the dominant tidal current and the long axis of the pockmark.

Figure 10. The wreck in the Witch's Hole.

Side scan sonar record (1987) showing the shipwreck within the Witch's Hole. Note also the unit pockmarks and the high reflectivity of the seabed within the pockmark.
Scale marks are 25m. apart.

Details of the wreck are unknown, however the dimensions and frame spacing suggest that it may be a Zulu-class wooden-hulled fishing vessel (L. Stule, 1987; Pers. Comm.). These operated from the north east coast of Scotland from the turn of the century until the 1930's or 1940's. Evidently it has lain in the Witch's Hole since before the 1975 survey and a date three or four decades earlier is not unreasonable.

The most intriguing question about this wreck is, how did it come to be in the Witch's Hole? The following possibilities must be considered:

1) The pockmark formed by scour around the wreck. This is highly improbably because the tidal currents are not very strong in this area. Furthermore the evidence of gas around and beneath this feature suggest that it was formed by gas escape.

2) The ship foundered as a result of its nets becoming entangled in something within the pockmark, possibly a glacial dropstone. This seems unlikely.

3) The ship sank and landed in this particular pockmark by pure chance. In this area pockmarks occupy approximately 6% of the seabed and of this only about 1 in 500 is active, consequently the probability of this happening by chance seems unreasonably low.

4) The ship lost buoyancy and sank as a result of a gas escape event. This would certainly explain why the wreck lies in an active pockmark, but the probability of the ship sailing over the pockmark just as a gas escape event

occurred seems remote.

CONCLUSIONS

The evidence considered here relates mainly to one small area in the North Sea, however there is no reason to suppose that it is atypical of shallow gas areas. The evidence presented supports the supposition that gas is actively migrating through the seabed and sub-seabed sediments. This process has continued for a considerable period of geological time, and there seems no reason to suppose that it will cease in the foreseeable future. The rate of migration is difficult to ascertain, however there is evidence to suggest a measurable movement of gas near the seabed on a time span of significance to offshore operations and structures.

This preliminary investigation has produced evidence, albeit somewhat limited, that variations in the extent of acoustic blanking may occur over a period of a few years. This suggests that shallow gas might encroach upon a site during the lifetime of a structure. Periodic gas emissions capable of lifting sediment into the water column could alter the dimensions of pockmarks thereby affecting pipelines. Evidence of significant gas migration at depths likely to result in blow-outs has not been brought to light, however migration of gas at these depths must occur if the shallower accumulations are to be fed.

It is concluded that shallow gas is dynamic even over the (geologically) short time periods that are of initerest to offshore operations. If this conclusion is upheld by the more detailed work that is planned, or from similar studies conducted elsewhere, then it would seem wise to conduct repeat surveys of sites that are to be occupied for more than a few years in order to monitor the position of gas accumulations.

The significance of the wreck in the Witch's Hole is open to conjecture, however the possibility that occasional gas escape events may cause the loss of buoyancy of a vessel cannot be ruled out.

ACKNOWLEDGEMENTS

The data utilised during the preparation of this paper was made available by Total Oil Marine p.l.c. and the British Geological Survey to whom I am deeply indebted. I particularly wish to thank Peter Dyson and Dave Long for their help, Pat Cowell and Neil Purvis for assistance with the illustrations. Some of the initial work reported in this paper was undertaken by Kasmadi bin Kaling and Diane Robson.

REFERENCES

Aquatronics (1975). Engineering geophysical survey in block 21/3. Report to Total Oil Marine Ltd. (unpublished).

Faber, E. and Stahl, W. (1984). Geochemical surface exploration for hydrocarbons in North Sea. AAPG (Bull), 68,363-386.

GECO (1989). Tectonic map of the North Sea. GECO Exploration Services and Alastair Beach Associates, Woking.

Hovland, M. and Judd, A.G. (1988) Pockmarks and seabed seepages: impact on geology, biology and the marine environment., Graham and Trotman, London.

McQuillin, R. and Fannin, N.G.T. (1979). Explaining the North Sea's lunar floor. New Scientist 83(1163),90-92.

McQuillin, R, Fannin, N.G.T. and Judd, A.G. (1979). IGS Pockmark Investigations 1974 - 1978. Report No. 98, Marine Geophysics Unit, Institute of Geological Sciences.

Sweeney, R.E. (1988). Petroleum-related hydrocarbon seepage in a Recent North Sea sediment. Chemical Geology, 71,53-64.

Stoker, M.S., Long, D., and Fyfe, J.A. (1985). A revised Quaternary stratigraphy for the central North Sea. Report 17(2), British Geological Survey.

Part II
Acquisition

HIGH RESOLUTION SEISMIC SOURCE SPECIFICATIONS APPLICATIONS AND MISAPPLICATIONS

C. R. Price, B.Sc., FGS

Geodetic International, Inc., Houston, Texas, USA

INTRODUCTION

Source specifications for high resolution seismic surveys have only recently been introduced into contract documents and this development is significant in terms of the data quality which can be achieved if the specifications are properly applied. Over the past ten years it has been traditional that site surveys, and other similar projects involving the use of digitally-recorded high resolution seismic data, have not been regarded very highly by the managements and exploration and production departments of the Client companies. Accordingly, it was often left to the survey department, if one existed, or an operations group, to acquire the high resolution data as quickly and as cheaply as possible, with minimal disruption to the "more important" operations in progress. Processing and interpretation of the data were viewed in the same way. As long as a basic "bright spot" detection exercise could be said to have been carried out, all the requirements and interests existing at that time were

Volume 25: Safety in Offshore Drilling, 53–73.

satisfied. Given these circumstances, it is not surprising that seismic source specifications were not even considered when planning high resolution surveys.

Finally, in 1990, it appears that a small proportion of the offshore operators now recognize that there is a requirement for acquisition of optimum quality high resolution seismic data, for the efficient detection of shallow gas hazards. This recognition has developed as offshore drilling activity has continued at a high level all over the world and the safety considerations have escalated in response to a number of well-publicized accidents and attendant public awareness. The importance of optimizing high resolution seismic data, in order to provide information vital for safety during offshore drilling, has been clearly demonstrated in numerous instances in the past. Seismic source specifications are fundamental to the improvement of high resolution data, and to the detection of gas hazards, and this paper demonstrates the effects which can be created by properly-applied and improperly-applied specifications. It is hoped that an awareness of these effects may encourage other operators, who currently have little regard for high resolution seismic data, to recognize the potential for effective hazard detection offered by the method and the risks involved if inadequate site surveys are carried out.

HISTORICAL REVIEW

In the late 1970's and early 1980's the source used by most high resolution seismic contractors was the sparker, or sparkarray.

This was because it was the cheapest and easiest source to operate at the time. The output pulse characteristics of the source, which determine the limits of resolution and penetration attainable using that source, were of no concern and were not usually specified. This was just as well, because a typical sparker pulse, shown in Figure 1, is rather difficult to provide specifications for, in the normally accepted terms.

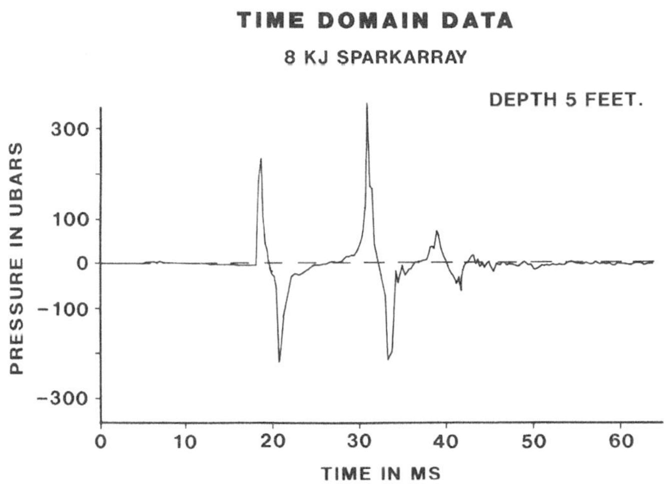

FIGURE 1

The normally accepted parameters for specifying source characteristics are those used in the exploration seismic industry and involve one expression to define signal strength and another to define signal shape.

In the time domain, **source strength** is quoted in pressure/distance units, which are normally bar-meters in the seismic industry. The micropascal (10^{-11} bar) may also be used in place of bars. The source strength measurement is made on the primary arrival in a signature and is either given as the peak strength or, more commonly, peak-to-peak (PTP) strength. This is indicated on Figure 1, but, given the sparker signature shown here, it is difficult to identify the primary arrival in order to make this measurement.

The **source shape** is represented in the time domain by the primary-to-bubble ratio (PBR), which is simply the PTP strength of the primary arrival divided by the PTP strength of the residual bubble pulse. Again, this is indicated on Figure 1, but is very difficult to determine in any meaningful way, given the sparker pulse characteristics.

The third element which is normally accepted as being critical when considering source specifications is the stability, or repeatability, of the pulse. This is not illustrated here, but as the sparker pulse typically changed quite radically from shot to shot, any quoted signature specifications based on a single pulse would be rendered quite meaningless.

Other sources were available in the early 1980's for high resolution data acquisition and, depending upon the contractor, the miniflexichoc, air guns, minisleeve exploder and water guns were all introduced as the years went by. This led to what could be considered a normal source specification for high resolution surveys in these days, an example of which, taken from an actual contract document issued in 1983, is shown below.

> *"This tool shall provide the reflection digital reading from a sparker, water gun, air gun, sleeve exploder or other*

proven and agreed to high resolution seismic energy source. The power and frequency spectrum must be similar to a single 80 cu. in. water gun or 10 to 15 kilojoule multi-tip sparker."

Apart from giving a choice of source which could be used, this introduces a technical note, equating a water gun signature with that of a sparker source, without actually quoting any source specifications. This is interesting, because if we compare such a water gun signature, shown in Figure 2, with the sparker signature shown in Figure 1, it is obvious that there are few similarities in the time domain, or in the frequency domain (Figure 3). Thus we can see that the "specification" provided was misleading, misapplied and of very little use in terms of ensuring acquisition of true, good quality high resolution seismic data.

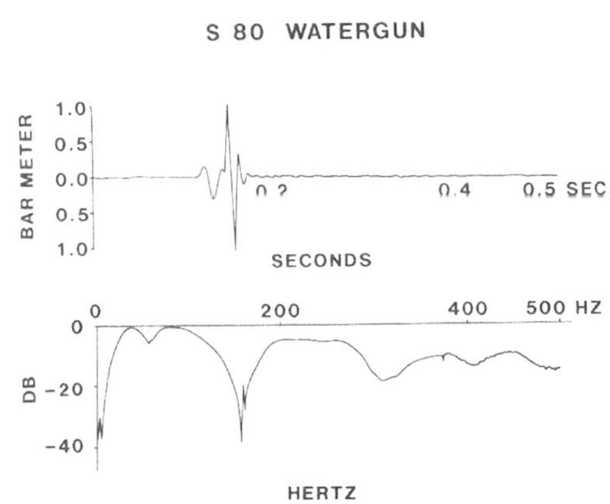

FIGURE 2

Perhaps the following source specification, also taken from an original contract document, was the best which could be issued at this time.

"The sound source shall be a high frequency source with high firing rate capabilities such as: miniairgun, watergun, minisleeve exploder, miniflexichoc or equivalent."

This specification is confined to a list of acceptable sources and it is obvious that whoever compiled this document either had no knowledge of, or no interest in, the signatures of each of the sources listed. A comparison of the signatures shown in Figures 1, 2 and 4 illustrates the radical differences between the performances of the sources deemed acceptable, which will determine the type and quality of the data acquired by each. The

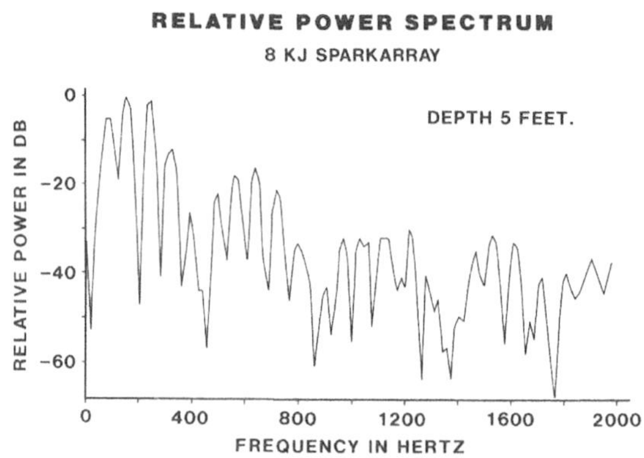

FIGURE 3

time domain plots clearly show differences in signal strength and shape, involving the primary arrivals, pre-cursor events, bubble pulses and other characteristics. In the frequency domain, even more significant differences are revealed, as shown by the comparison between sparker and minisleeve exploder frequency plots, in Figure 5. The -6dB and -20dB relative power levels are included on these plots to facilitate comparison of the frequency content of the signatures at these levels.

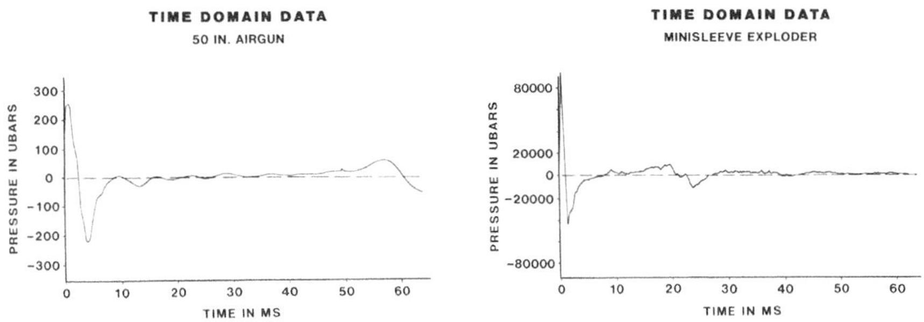

FIGURE 4

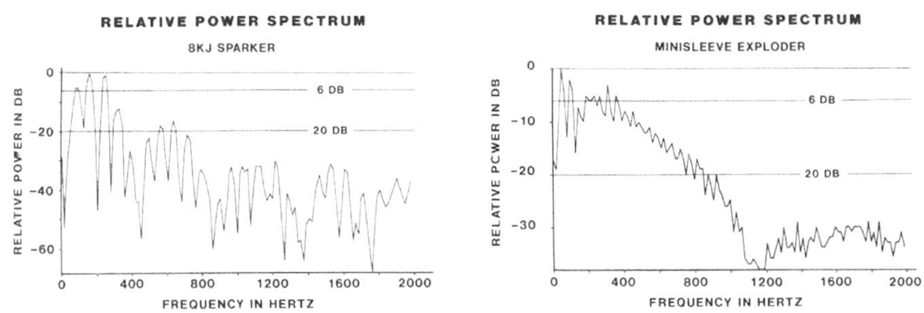

FIGURE 5

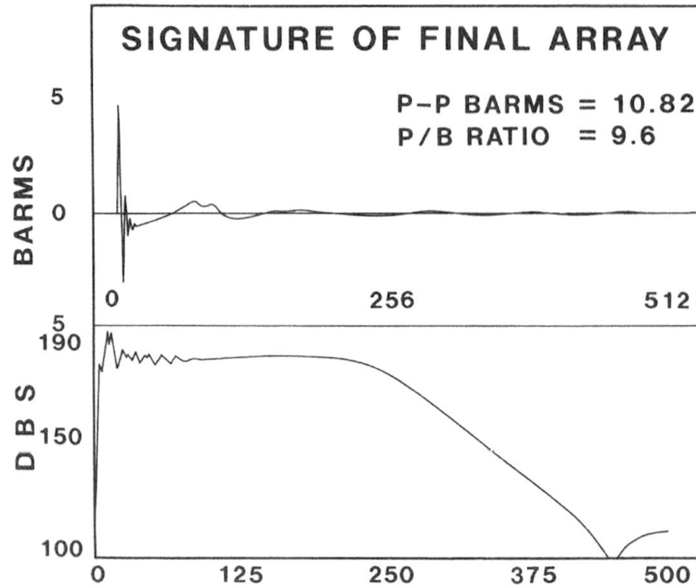

FIGURE 6a

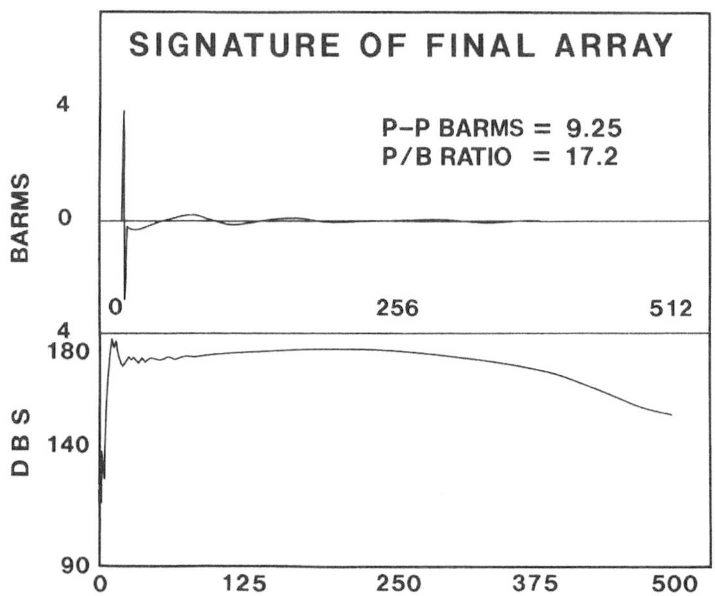

FIGURE 6b

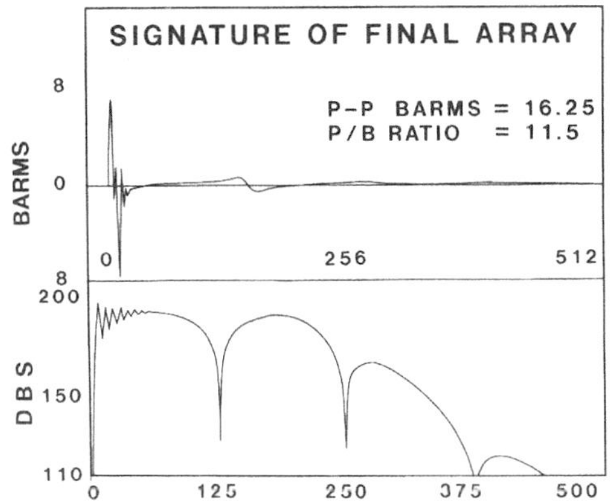

FIGURE 7

The introduction of the minisleeve exploder as a high resolution source which was actually capable of true high resolution (0.25 millisecond sample rate) data acquisition, was instrumental in bringing the source performance characteristics to the attention of the survey planners. This source illustrated that most of the other systems being used were not actually very suitable for high resolution data acquisition, and that the range of pulse shapes, strengths and signature characteristics being utilized was huge. More importantly, it also demonstrated very clearly that the signature of the source utilized had a direct bearing on the quality of the recorded data and the ability to differentiate and detect anomalies and hazards in the data.

An example of the specification which began to be issued at this time is shown below.

"A source delivering a clear and repetitive signal with high energy and a wide frequency spectrum will be operated......

Any of the following sources will be eligible:

- *Single air gun (100 cu.in. for example) or gun pattern;*
- *Miniflexichoc;*
- *Sparker (about 10k Joules) minimum of 18 candles (2 x 9); multitip candle option shall be proposed;*
- *Water gun (200 or 400 cu. in.)"*

This shows a specified energy level (high), frequency spectrum (wide), signal stability (repetitive) and signal shape (clear). The fact that none of these specifications can be said to be very rigorous, or applied correctly by a knowledgeable or interested geophysicist producing a tender document, is illustrated by the second part of this particular source specification. The list of eligible sources contains all those systems with wildly varying source characteristics which are illustrated in Figures 1 through 5. This entirely negates the first part of the specification and results in a misleading and misapplied source characterization.

CURRENT STATUS

Since 1987, considerable technological developments have taken place in the conventional seismic acquisition industry and, simultaneously, significant attitude adjustments have occurred within the offshore exploration industry. The attitude changes are linked primarily to the price of oil and to exploitation costs since the last major industry downturn in 1986, which have created the necessity for the most benefit to be squeezed out of every dollar spent during exploration and production. The secondary cause of attitude changes has been the increased awareness of safety considerations during offshore drilling and the potential risks involved if sufficient precautions are not taken. This has had significant effects on the high resolution seismic industry, as the awareness level has grown among many Client companies that properly acquired high resolution data are of fundamental importance as a basis for safety during offshore drilling and that the technology exists to allow such data to be realized.

Source specifications did not immediately become modified in any major way, as shown by the 1987 example below, where Air/Water guns are still equated with a sparker source, but a new element was introduced.

> *Air gun(s) or Water gun(s) : Either a single gun or tuned array may be used. Source repetition period must be not greater than 3 seconds....*
> *Sparker : A sparker seismic source may be proposed with a minimum power of 15kJ. The source must be capable of firing at least every 4 seconds...."*

The introduction of the possibility of using "tuned arrays" of air guns, rather than single guns, increases the scope of the source specifications dramatically. Contractors offered high resolution arrays, based on air guns or water guns, using the research and development carried out by the conventional exploration seismic industry as the basis for their array designs. In every case the arrays were designed to improve the source characteristics we have already seen being specified in the past, **source strength, signature shape and stability,** in order to optimize the acquired data. This was achieved by varying the number of guns, gun volumes, gun spacings and many other parameters; including gun type when new, improved air gun designs came on the market. With this number and variety of parameters which could be modified in any one source, it became critical that meaningful source specifications could be provided, to ensure good, consistent data could be acquired.

ARRAY SPECIFICATIONS

In any given array, the following factors are those which may be easily modified and which may have significant impact on the source signature produced.

Gun Type
Number of Guns
Individual Gun Volumes
Total Array Volume
Gun Separation
Gun Synchronization
Firing Pressure
Port Depths
Port Areas
Port-closure Pressure

To specify all the above would be ridiculous and unnecessary, as the considerations which are important in order to realize optimum quality of recorded data remain exactly as they were when the sparker was the chosen source - **source strength, signature shape and repeatability.** It becomes necessary, therefore, to be able to isolate the parameters by which these elements can be precisely defined in any source specification and leave selection of all other array design features up to the contractor. Also, it must be considered that, in order to optimize data to the full extent allowed by the new source technology in a cost-effective manner, the specification parameters must be much more rigorously detailed than previously. This leads to increasing examination and analysis of source signatures, by means of field experimentation and computer simulations, and to acceptance or rejection of a particular array based on submitted source signatures. If this is to become standard procedure, the source characteristics issued in tender documents must be precisely defined and specified, or misleading data will be submitted and mistakes may be made in source selection and rejection. More significantly, unless the specifications are properly defined, a relatively poor source may be contractually acceptable, resulting in less than optimum data and a decrease in efficiency and cost-effectiveness of the survey. This will impact the safety aspect of the survey, as poor data may result in hazards remaining undetected, as has happened in the past.

An example of a recent, better defined source array specification is shown below.

> *"....the signature of the source shall be measured in deep water. Throughout the work the source will be capable of producing a stable outgoing signal, characterized by a <u>peak-</u>*

to-peak amplitude of 5 barmeters and a primary to bubble ratio of 10 to 1, both measured using a 1000Hz/12dB filter...."

In this case, each of the important elements, stability, strength and shape, is precisely defined and also related to a new and very important parameter - the frequency bandwidth or hi-cut filter used to record or compute the time-domain specifications. Both the peak-to-peak strength and the primary-to-bubble ratio of a signature in the time domain are strongly dependent upon the bandwidth in use. If the PTP and PBR of a source are specified, then the bandwidth used to attain these values must also be specified.

SLEEVE GUN ARRAYS

To illustrate the effects of band-limiting a signature, and several other aspects of array specification design, a series of results from computer simulation exercises carried out for a theoretical Sleeve Gun Array are presented below. The computer program used for these simulations was developed by Dr. R. Cotton of Halliburton Geophysical Services, Inc., and the assistance of Dr. Cotton and HGS in the production of these simulated results is gratefully acknowledged. The simulation program has been exhaustively tested against field recorded results and proven to be highly accurate under a wide variety of production conditions. Figures 6 a) & b) show the effect of different high cut filters being applied to the signature data and demonstrate that the peak-to-peak strengths and primary-to-bubble ratios of sources are strongly dependent on the frequency bandwidths in use.

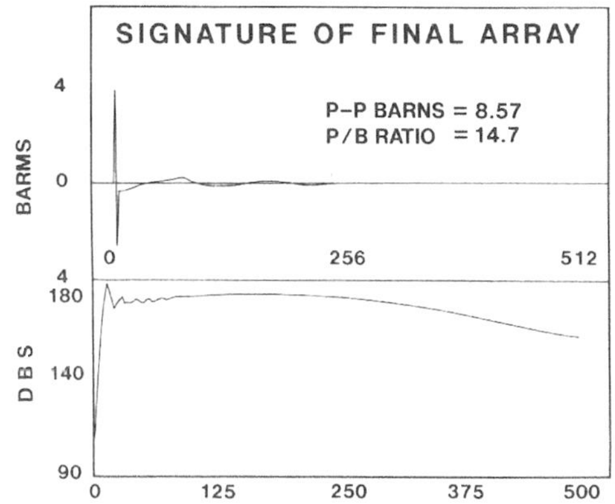

FIGURE 8

Therefore, if time-domain specifications are to be used, the bandwidth must always be specified. In Figures 6 a) and b) the peak-to-peak strength figures do not show the same remarkable variation as the primary-to-bubble ratio figures, because the simulation program is set up to model the venting of energy through the sea surface, which is characteristic of a shallow-towed source. This introduces a slight anomaly into the primary energy measurements.

The second critical element which should always be included in a source specification is a delineation of the required frequency range, which must be available at a stated relative power level. Analysis of this element can only be carried out by examining the pulse signature characteristics in the frequency domain. Figure 7 shows a signature which, in the time domain, may be acceptable to many Clients. It has high power (16.25 bm), good pulse shape (PBR 11.5:1) and is stable (not shown). However, the frequency domain plot clearly shows that this source would be entirely unsuitable for high resolution acquisition, as it lacks high frequency (>250Hz) content and has deep notches in the frequency spectrum at critical bandwidths (125Hz).

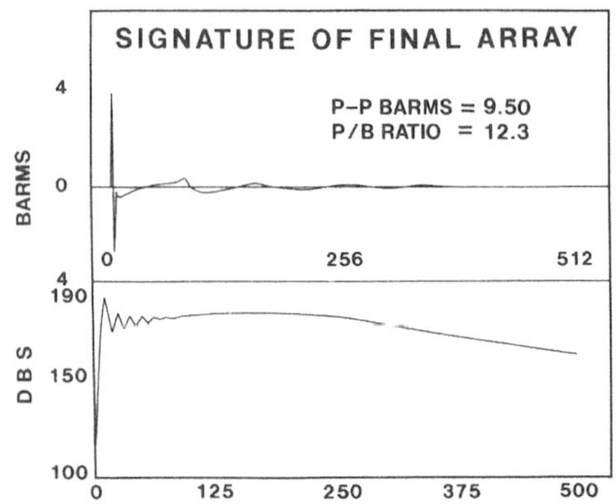

FIGURE 9

Figure 8 also shows an acceptable time domain signature. The source strength is lower than in Figure 7 (8.57 bm), but adequate for high resolution acquisition, and the pulse shape is excellent (PBR 14.7:1). This pulse is also very stable. Although the time domain characteristics shown in Figures 7 and 8 are similar, the frequency domain plots show remarkable differences. In Figure 8, high frequencies are present at much greater source levels to in excess of 400 Hz and the detrimental notches are absent from the spectrum. This souce would be suitable for many Clients and would perform well.

Thus, the elements which are critical to the specification of a source performance adequately, completely and accurately, have been defined. These are summarized below.

	i) STABILITY	
	ii) SIGNAL STRENGTH (PTP)	
TIME DOMAIN		BANDWIDTH
	iii) SIGNAL SHAPE (PBR)	
	iv) BANDWIDTH REQUIRED (min. & max. Hz)	
FREQUENCY DOMAIN		
	v) RELATIVE POWER LEVEL (-dB's)	

Using these five fundamental elements, a precisely defined, properly constructed and correctly applied specification can be produced which will ensure that a source capable of recording optimized high resolution data is supplied by a contractor. An example of such a specification is shown below.

".....the source will be capable of producing a stable outgoing signal, characterized by a peak-to-peak amplitude of 6.0 barmeters using a 750Hz/70dB filter. The primary to bubble ratio peak-to-peak shall be better than 15:1. The source signature shall have an amplitude spectrum with as a minimum -6dB points at 10Hz and 300Hz."

This is the starting point for acquisition of data of the standard necessary for efficient hazard detection prior to offshore drilling.

It is important to note that, provided the above specifications can be met, the source itself is immaterial. It could consist of one gun or many guns and, in an array, the gun spacing, gun depths etc., would all be immaterial; provided the specifications are reasonable (as above), can be met by the source supplied and can be proven to be met, by means of source monitoring regulated by the Client. In practice, however, it is found that in order to meet well-defined specifications such as those quoted above, very precise array designs must be employed. The series of examples given below shows the effects of varying gun depths, gun spacing and gun volumes in a array. From these examples it can be seen that these, and other elements, are critical to array performance and that the signal characteristics can be modified radically by slight changes in array design.

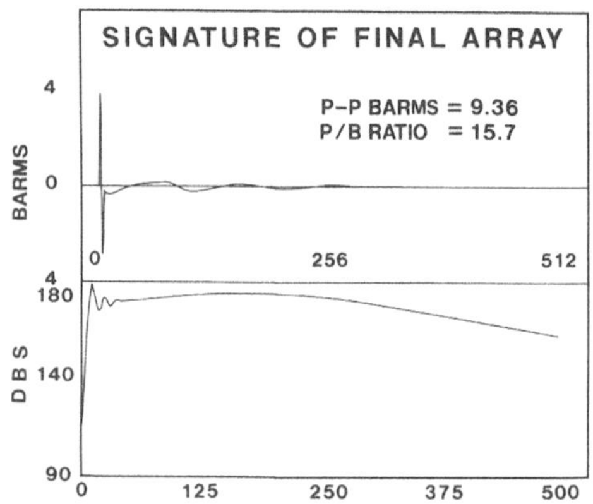

FIGURE 10

Figure 9 shows the signature from the same array which produced the output shown in Figure 8, a vertically staggered array of mixed-volume guns, but with the depth of the largest gun in the array increased by 20 centimeters. The peak power has increased, to 9.50 bm from 8.57 bm, as less energy is vented through the sea surface from the shallow-towed array. The primary-to-bubble ratio decreases, to 12.3:1 from 14.7:1, due to the production of a larger bubble, with more residual energy, and less coalescing of the bubbles from all the guns in the array.

By simply reducing the volume of the deepest gun in the array, Figure 10, the peak power can be maintained, at 9.36 bm, while increasing the primary-to-bubble ratio significantly, to 15.7:1 from 12.3:1.

Figure 11 shows that the effects of reconfiguring the array to achieve the optimum cluster effect, by equi-spacing the gun ports at a critical distance from each other and below the sea surface. The gun volumes are not changed from the previous examples, but the geometrical arrangement has been modified. This has achieved a modest increase in peak power, to 10.25 bm, and a significant increase in primary-to-bubble ratio, from 15.7:1 to 34.7:1.

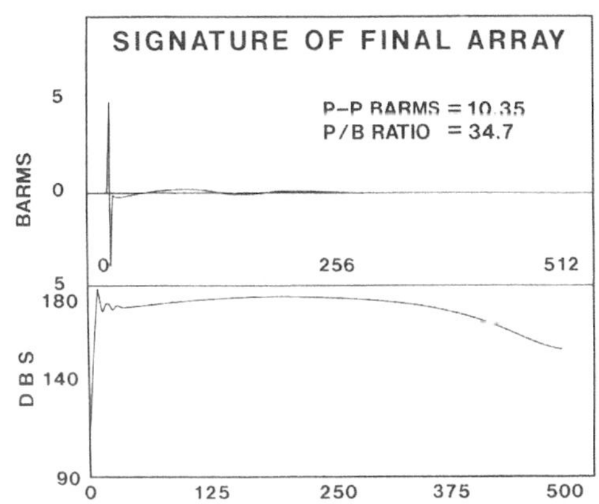

FIGURE 11

The effect of significantly increasing the total array volume, in this case by 150%, by using the same number of larger guns, is shown in Figure 12. It can be seen that an increase in peak power of only 20% results, from 10.35 bm to 12.13 bm, while the primary-to-bubble ratio remains constant. This is not surprising when considering that the peak power of an air gun is proportional to the cube root of its volume. Thus, doubling the

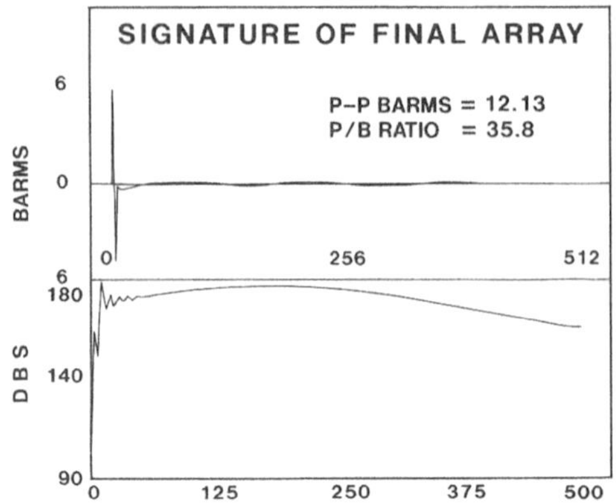

FIGURE 12

volume of a single gun will only increase its power output by 26%. However, by increasing the number of guns in an array, but keeping the total array volume constant, an increase in peak strength of up to 65% can be achieved without the use of any additional air. The number of guns employed in an array therefore becomes critical when designing a source for high resolution acquisition, given specifications which demand relatively high power output and high primary-to-bubble ratios.

SUMMARY

There are many elements involved in the efficient and cost-effective detection of shallow gas hazards offshore. Processing and interpretation of data are two major areas of concern which must be addressed, but the field acquisition parameters are a fundamental starting point for the production of the necessary high quality data, without which no interpretation can be successfully completed. Source specifications are one of the elements of the acquisition parameters which can have an enormous impact on the quality and type of data recorded. If properly applied, source specifications can help optimize the data acquired and significantly increase the efficiency of identification and detection of shallow gas hazards offshore. If misapplied, the source specifications can contribute to poor data acquisition and result in hazards remaining undetected and the safety of rigs during offshore drilling being significantly reduced.

ENHANCED SEISMIC METHODS FOR MARINE HAZARD SURVEYS

P. Newman

Consultant, formerly Horizon Exploration Ltd.,

Swanley, Kent, England.

ABSTRACT

Using results obtained from existing 3-D marine seismic surveys acquired with a compact recording geometry we show how true-amplitude migration processing can yield remarkably well-defined images of the shallow geology in plan view. Examples of such images are presented showing features of small physical dimensions, as well as a number of responses that are presumed to represent gas hazards. The pictorial quality of intensity-modulated time slice displays is particularly striking. In many cases small yet potentially important targets could remain unsampled by the customary grid of 2-D survey lines, or would be virtually uninterpretable without the benefit of 3-D migration. Even when 3-D methods have been applied, conventional seismic cross sections are inappropriate for displaying responses resulting from fractures or channels that have a small spatial footprint. Commonly, these will remain unnoticed or be dismissed as noise in a vertical seismic profile. The Short-Offset field technique that we propose for this work can be economically applied to hazard surveys. Results suggest that the temporal bandwidth commonly specified for shallow profiling can probably be relaxed when areal

Volume 25: Safety in Offshore Drilling, 75–86.

data recording methods are applied. Routine application would reasonably be expected to reduce drilling costs whilst significantly improving safety.

INTRODUCTION

Seismic techniques as applied to hazard surveys are essentially the same as those used for deep exploration, differing primarily in scale. The seismic source, record length, and sample rates both spatial and temporal, are adjusted to suit the shallower depths of investigation and the broader signal bandwidth that is expected to apply. Recording instrumentation, processing, and final display methods also follow those employed for conventional surveys, resulting in a grid of intersecting 2-dimensional vertical profiles.

An immediate question concerns the suitability of the conventional seismic cross section display for site survey applications. Of principle interest in these studies is the detection of amplitude anomalies and determination of their spatial extent. A further aim is to establish the competence or otherwise of shallow formations, as may be indicated for instance by evidence of faults or fractures. We hope to demonstrate that these objectives are better realised using sequences of horizontal seismic profiles displayed in variable intensity mode.

In order to generate horizontal profiles it is of course necessary to acquire seismic data over a closely sampled area, such as is regularly accomplished for conventional 3-D seismic surveys. For shallow objectives, however, long streamers and large fold of common mid-point (CMP) stack are unnecessary and much of the complexity associated with 3-D marine seismic surveys is thus avoided. Indeed, the hazard survey appears to be an ideal application for the so-called Short-Offset 3-D method described by Newman (1984, 1985, 1988, 1989.).

A key requirement of the Short-Offset method is to approximate normal-incidence reflection responses, as nearly as is practicable, by operating with the minimum source-receiver separation that is consistent with tow noise and recording dynamic range considerations. Data that are recorded in this way are well conditioned for predictive deconvolution and for a modified Kirchhoff migration process that simultaneously compensates the amplitude decay that accompanies geometrical spreading. The result is an intrinsically simple true-amplitude processing sequence that is well suited for hazard interpretation. The compact nature of the source-receiver configuration provides important operational advantages, making this a potentially efficient and economical procedure, particularly attractive for site survey work where the area to be examined is comparatively small.

We might reasonably anticipate some difficulty in reconciling the bandwidth and sample rates that commonly apply to shallow survey data with design criteria for effective cross-line migration operators. A further perceived problem might be that of maintaining line separation within tolerances that are acceptable in the context of the elevated frequencies employed, and given the likely errors in absolute position fixing. In practice, small-scale features appear to be well resolved in horizontal profiles at relatively low seismic frequencies, such that sample rates in both time and space can be relaxed when these methods are employed.

A number of curiosities relate to time-slice displays which in combination make these attractive for hazard interpretation. Regardless of the frequency bandwidth of the seismic waveforms that are sectioned to produce a time slice, the horizontal wavenumber spectrum can extend from zero to infinity depending on the attitude of reflectors. A simple fault in a plane, horizontal reflector results in a stepwise discontinuity in display intensity in x-y space. Moreover, the change in image intensity across such a discontinuity is influenced by a combination of trace amplitude and the phase delay occasioned by the fault. In vertical profile a fault is evident as a break in continuity that may be very slight and highly localised. In time slice, the same fault will exhibit lateral continuity and is easily recognised.

Perhaps the most important property of the time slice, however, is the way that it emulates a geological map, an aerial photograph, or perhaps a sidescan sonar display. Interpretation is aided by the powerful attribute of physical plausibility.

Several of the examples from field data that follow are thought to illustrate responses from gas bearing rock, although none of these has been verified by the drill. We particularly want to convey the pictorial qualities of the variable-intensity time slice display, and to indicate the sensitivity of the method to subtle changes in seismic response that are simply not evident, nor capable of interpretation, in vertical profile.

EXAMPLES FROM FIELD DATA

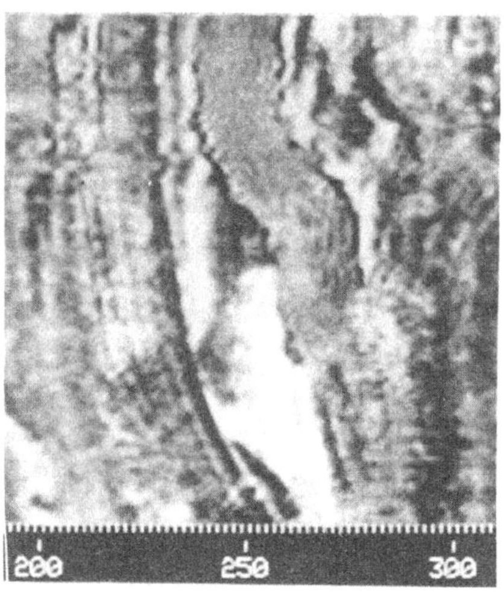

Figure 1. Irish Sea time slice at 164 milliseconds. Area shown is 3km by 3km.

The first example comes from the initial test of the Short-Offset method in the Irish Sea, described by Newman (1985). Figure 1 is a portion of the time slice at 164 milliseconds and covers an area of 3km by 3km. The scale marks correspond to the shot point interval of 25 metres. This survey was recorded with a line spacing of 25 metres, source-receiver offset of 55 metres between array centres, and was processed single fold. Not intended as a high resolution shallow survey, the source array comprised a total of 16 S80 water guns in two strings separated by 25 metres and having an in-line extent of 50 metres. Processing was at 4 millisecond sample rate and included an adjacent trace sum of recorded 25 metre groups and 10-60 Hertz bandpass filter. The display is simply the signed sample value of each seismic trace at

164 milliseconds converted to a grey scale. All of the time slice illustrations that follow were photographed directly from a GeoQuest IES screen.

Of particular interest in the Triassic setting of figure 1 is the channel, approximately 500 metres wide, running in a roughly N-S direction. Variations in image intensity within the banks of the channel give the impression of an undulating surface such as might be expected to result from sediment transported by the ancient river. In contrast with this variability, notice how consistent is the image to the west of the channel, presumably representing a flood plain. This is bounded further to the west by steeply dipping and eroded subcrop.

Figure 2 has greater relevance to the shallow gas problem. This is a time slice at 476 milliseconds from a survey conducted in the Danish sector of the North Sea, which again was targeted at deep objectives. The area shown is about 13km by 8.5km. Line spacing and shot interval were both 25 metres

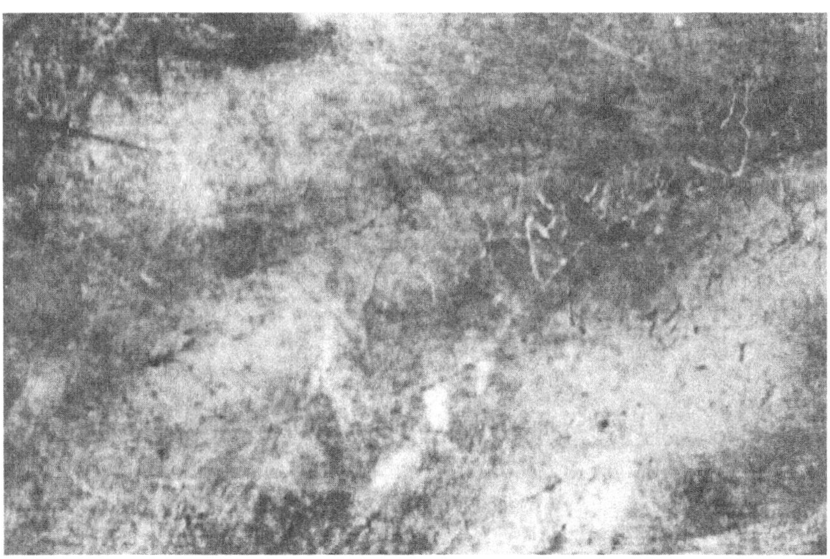

Figure 2. Time slice at 476 milliseconds, Danish North Sea, area 13km by 8.5km. Irregular features are expected to represent gas-charged ice scour channels.

although processing was accomplished at 12.5 metre depth point spacing, this being achieved by interpolation in the cross line case. Bandwidth was restricted to 80 Hertz at this reflection time, which corresponds to an interval within the Quaternary. There can be little doubt that the seemingly haphazard lineaments result from ice scour action, or that the localised bright spots indicate the presence of gas. In total, an interval of about 100 milliseconds two-way time is influenced in this way with many of the channels being interconnected. The probability of intercepting one or more of these with a borehole must be unacceptably high.

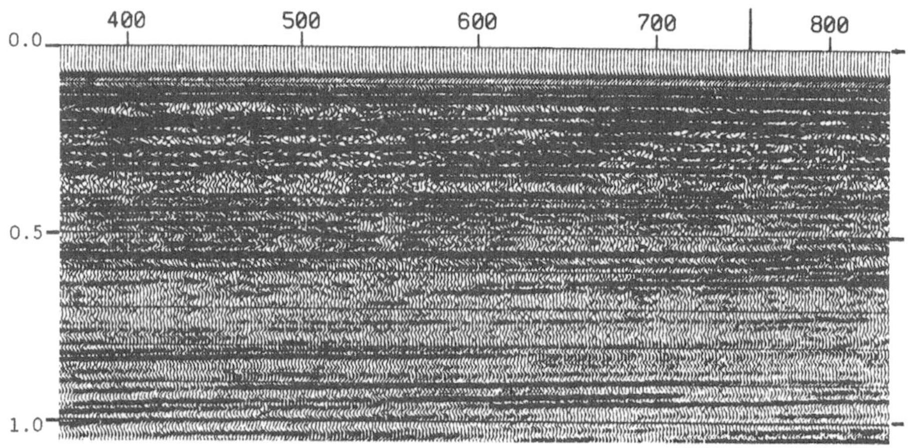

Figure 3. Enlarged portion from middle of figure 2 central line. Length of line shown is approximately 6km.

Although we no longer have ready access to intermediate unmigrated results for confirmation, it seems unlikely that bright spots of small physical dimensions as in figure 2 would be identified without the benefit of 3-D migration. In the 3-D migrated conventional vertical cross sections they are frequently seen to influence only one or two traces, and would be readily discounted as noise. For example, figure 3 shows an enlarged shallow portion of the central line, from the middle of the figure 2 survey area. A number of brief transient responses can be seen at times of about 0.5 seconds which would be expected to tie with the presumed ice scour channels, but which are uninterpretable on the basis of the vertical profile.

The remaining examples are taken from a shallow survey conducted adjacent to the Ekofisk platform complex, intended to investigate the effects of subsidence that has accompanied depletion of the reservoir. The total survey area in this case was nominally 4km by 2km. Data were recorded at 1 millisecond sample rate from a single 400 cubic inch water gun source. Shot interval and line spacing were both 10 metres whilst the group length was 5 metres. The processing sequence included 6-fold CMP stack and a bandpass filter having an upper limit of 100 Hertz. Figure 4 shows line 100, close to the centre of the surveyed area.

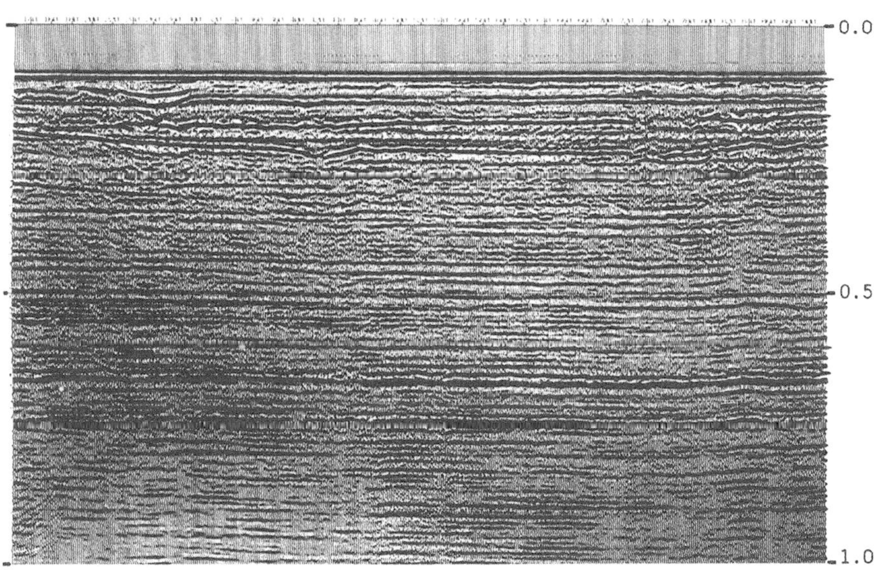

Figure 4. 3-D migrated central line from Ekofisk survey.

The first time slice that we show nicely demonstrates the sensitivity of these displays to targets that have small dimensions compared with seismic wavelengths. At a two-way time of 100 milliseconds, the time slice of figure 5 intersects the sea bed and clearly shows a number of pipeline trenches that converge upon the platform complex located approximately 150 metres from the

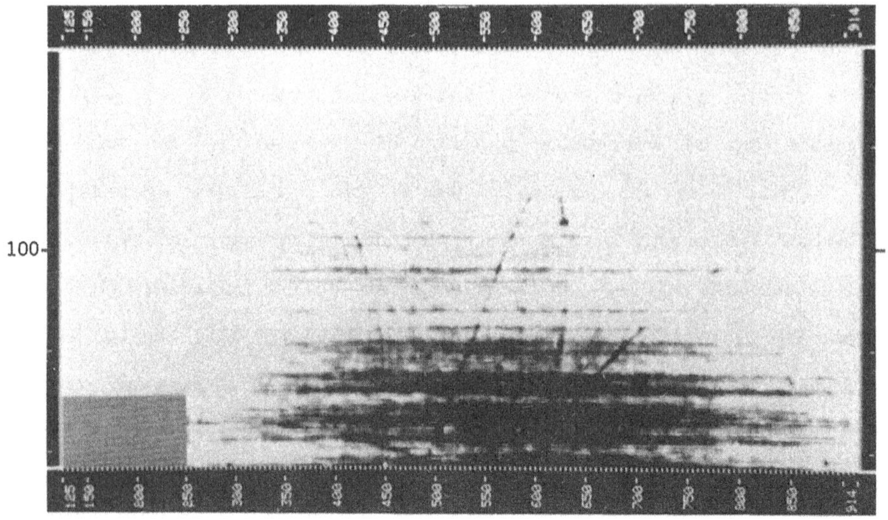

Figure 5. Time slice at 100 milliseconds, Ekofisk survey. Curvilinear features correspond to pipeline trenches.

lower edge of the displayed area. Trenches were dredged to a depth of 1.5 metres, 4 metres wide. The dark, roughly semicircular region at the bottom of the figure corresponds to the subsidence bowl at the sea bed, which results in a reflection time differential of about 3 milliseconds over the two-kilometre distance from top to bottom of the display. The horizontal striping in the figure is caused by small errors (less than 1 milliseconds) in correcting for tidal variations during the course of the survey.

A further intriguing example of the definition that can be achieved with moderately high resolution 3-D seismic methods is provided by the time slice of figure 6, at 237 milliseconds. This clearly shows the meandering course of an ancient stream within the confines of a broader channel. The area shown here is approximately 1km by 1km. Scale marks at the top of the figure are at 25 metre intervals, indicating that the width of the stream is less than this and possibly little more than 10 metres. The main channel can be easily recognised in the vertical profile of figure 4, approximately one third of the line length from the left hand side of the section.

A different type of response, which is presumed to result from gas accumulations, is shown in figure 7 from a time of 357 milliseconds. Here we observe anomalies that are not characterised by particularly strong amplitudes. Rather, it is the shape of the anomaly in plan that sets it apart from more familiar geological features. These amorphous blobs stand out because the phase of the response differs from that of the surrounding rock. It is unlikely that such phenomena would be properly identified in conventional seismic section. Certainly, their form is such that 3-D migration is needed in order to image them correctly.

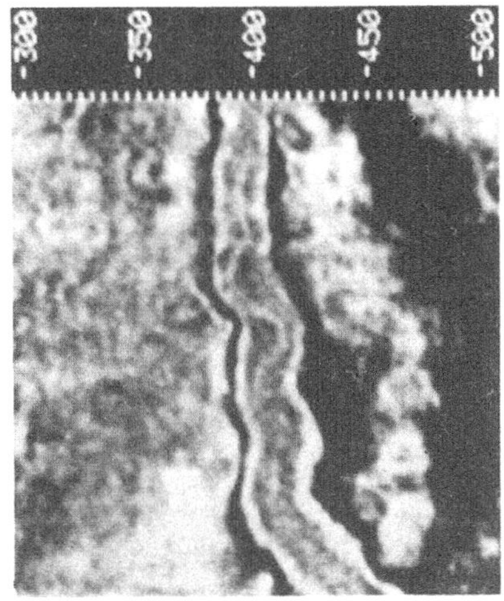

Figure 6. Meandering stream at 237 milliseconds. Area 1km square.

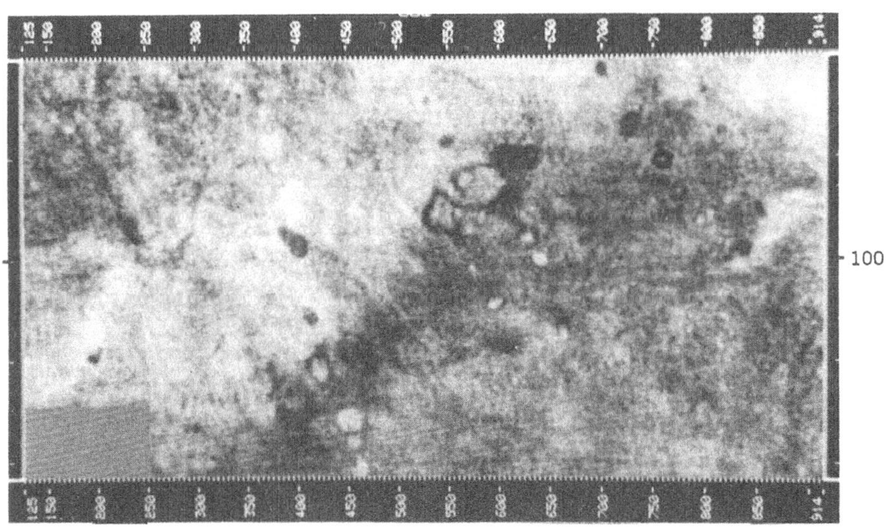

Figure 7. Time slice at 357 milliseconds, Ekofisk.

A final example of amplitude anomalies in plan view is provided by figure 8, a time slice from 603 milliseconds. In this case the anomalies are quite small in extent, but appear to be distributed along a number of linear trends that most probably correspond to fractures, although it is interesting to note that there appears to be no vertical displacement as would be shown by a change in grey level across the alignments.

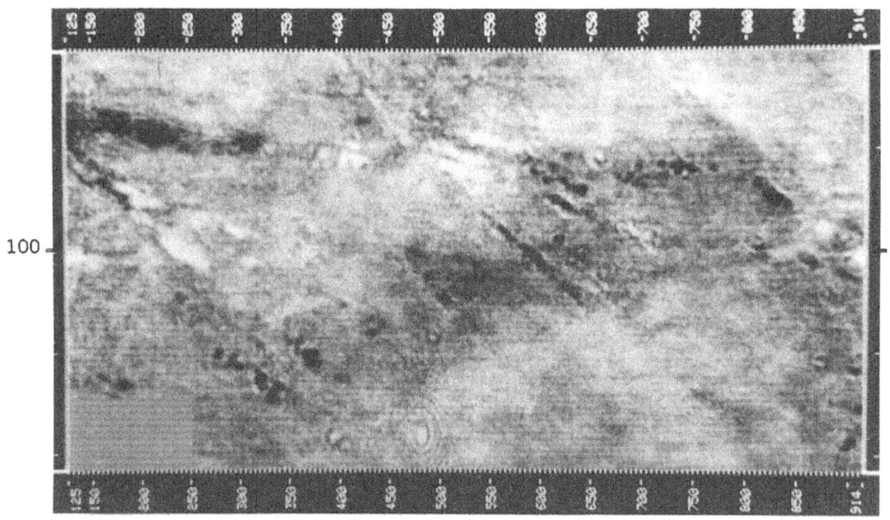

Figure 8. Time slice at 603 milliseconds, Ekofisk survey.

A surprising aspect of our studies of shallow time slices has been the rapidity of the changes that are observed, virtually from millisecond to millisecond which loosely corresponds to a metre by metre increment in depth. It is very easy to forget that the display is actually sampling seismic wavelets with an equivalent duration of perhaps a few tens of metres, and that the time slice display is in effect a composite of responses originating within a similar range of depths. To some extent this is due to the focussing action of the migration process; the time and velocity parameters that are defined for a Kirchhoff operator are appropriate to a unique time instant and not to all samples of a wavelet. The meandering stream of figure 6, for instance, is scarcely visible at 235 or 239 milliseconds. As in the optical analogy, the actual depth of field is determined by a combination of the aperture and distance to the focal point.

An interesting corollary to this is that at early reflection times, the principal effect of errors in migration velocity may be to focus some different part of the seismic wavelet rather than to degrade the image as a whole in plan view. Critical focussing effects must clearly become progressively less important with increasing reflection time.

CONCLUSIONS

A number of Short-Offset 3-D seismic surveys shot for relatively deep objectives but with short streamers and small source-receiver offset distances have produced images from shallow depths having a striking visual resemblance to monochromatic aerial photographs. The simplicity of the acquisition and processing procedures needed to achieve such results make this a particularly attractive technique for hazard survey applications. The areal survey method guarantees that no acoustic anomaly will be missed due to spatial sampling deficiencies, whereas there are strong indications that sampling density criteria can be relaxed vis a vis conventional hazard survey methods.

ACKNOWLEDGEMENTS

Many individuals within Horizon Exploration Ltd (now known as Simon-Horizon Ltd) made substantial contributions to the work reported here. The author is indebted to the Danish Underground Consortium for the example of figure 2, and to Phillips Petroleum Company of Norway and the Licence 018 group of companies for the Ekofisk examples.

REFERENCES

Newman, P. (1984). Marine Acquisition Techniques for Precision 3-D Surveys. Paper presented at the 46th Annual EAEG Meeting, London.

Newman, P. (1985). Short-Offset 3-D: A Case History. Paper presented at the 47th Annual EAEG Meeting, Budapest.

Newman, P. (1988). Cost-Effective 3-D Seismic Survey During a North Sea Winter. Expanded abstracts, 58th SEG Meeting , Anaheim, p.456-459.

Newman, P. (1989). Short-Offset 3-D Marine Seismic Applications. Expanded Abstracts, 59th SEG Meeting, Dallas, p.613-615.

FIELD QUALITY CONTROL

OF

MARINE DATA ACQUISITION - PAST, PRESENT, FUTURE

S. G. Middleton MSc

Hydrosearch Associates Ltd, Chandler House, Knaphill, Woking, Surrey, UK

ABSTRACT

Effective quality control of offshore survey operations has traditionally required the quality control representative to have a broad technical knowledge in a range of areas plus sound operational and commercial abilities.

Volume 25: Safety in Offshore Drilling, 87–100.

During the last three years however, there has been an increasing emphasis placed on new technical developments, many of which are based on the application of micro-computer technology. Specific examples of this are the use of on-board processing and the wider application of portable and laptop computer systems in a range of offshore operations.

In the short term, this has placed an increasing load on the offshore representative - both in terms of work load, and the requirement to be both computer numerate and to have a basic knowledge of seismic processing. The benefits of such trends have already been significant and promise to provide further benefits as technology advances further.

The paper will review current quality control procedures and comment on the impact which recent technological advances have had. Some ideas on likely development directions over the next three years will also be presented.

INTRODUCTION

Shallow Gas determination, prior to drilling, is carried out using 'Engineering Seismic' survey methods.

Seismic acquisition data quality is fundamental in achieving this objective as both the data processing methods and interpretation accuracy are subject to the quality of the data obtained.

The recent application of workstation based data analysis and interpretation has clearly demonstrated the importance of acquisition data quality on the ability to determine the risks of gas with a high level of confidence.

The following pages summarise the 5 primary issues that influence acquisition data QC, suggests future trends and considers their potential impact.

The primary issues covered in this paper are:

A. Ccismic

B. Positioning

C. QC Data Analysis

D. Communications

E. Personnel

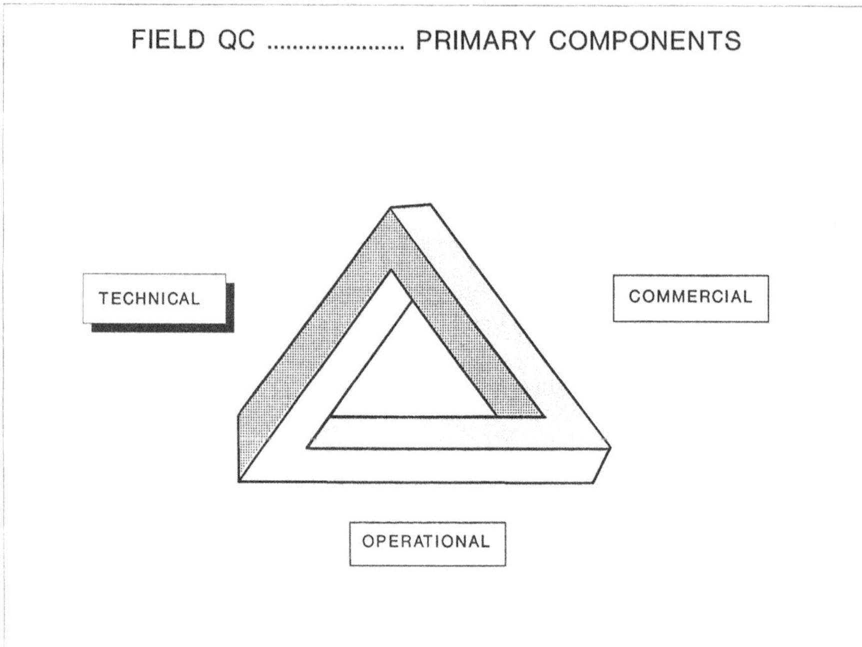

FIGURE 1

The often complex influences of Technical, Commercial and Operational constraints on the acquisition of high quality data require careful management by the QC supervisor in order to achieve a successful seismic survey data set.

The following pages only consider the Technical aspects of QC and assume that no Commercial or Operational restrictions are imposed in the achievement of optimum data quality.

It is noted however, that such restrictions are frequently responsible for significant compromise in data coverage and quality during routine rig site surveys.

Traditional 'Engineering Seismic' surveys utilise a number of Interfaced and Computer based systems (shown under '2' and '3' in

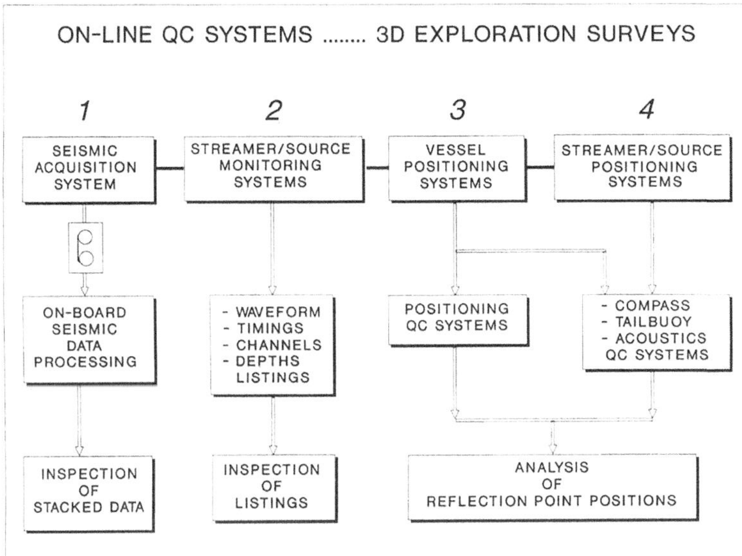

FIGURE 2

the figure) as the primary methods of monitoring and analysing the relevant data strings for each critical parameter of the survey.

These data strings usually provide, as output for QC, only line-by-line analyses in the form of separate listings of parameter values with Mean/Average/S.D type Histograms as line summaries.

'Exploration 3D' Seismics have adopted, in addition to the above, Onboard Seismic Processing ('1') and special Interfaced Streamer/Source positioning ('4') systems as standard techniques in order to measure the data quality in terms of Stacked Data and Reflection Point Positions respectively. This is because current requirements demand that 3D data sets achieve and maintain consistent data quality in both the Time (Temporal) and Space (Spacial) during acquisition.

ONBOARD PROCESSING SYSTEMS .... KEY COMPARISONS

	ADVANCE *MICROMAX*	ODEGAARD *RAMESSES*
HARDWARE PLATFORM	386/AP	MICROVAX
ONLINE PROCESSING		*
OFFLINE PROCESSING	*	*
REAL TIME QC OF RAW DATA		*
INSTRUMENT TESTS	*	
CAPACITY LIMITS	No OF CDP's RECORD LENGTH SAMPLE RATE HARD DISC SIZE	CYCLE TIME No OF CHANNELS
ACQUISITION VS PROCESSING RATES	1:2 APPROX	1:1 APPROX
UP-GRADING	YES	YES

FIGURE 3 HYDROSEARCH

These comparisons are intended to show the overall capabilities of 2 commercially available Onboard Processing systems.

Other on-line and off-line systems are in current use and future developments in computer hardware and software will undoubtedly bring further abilities to Onboard Processing.

The current difficulties in using Onboard Processing systems to effectively field QC the data are due to time and data storage limitations and a shortage of personnel skills/knowledge rather than limitations of the technology. Further developments and requirements should take full consideration of these points.

The ability of processing systems to test, analyse and correct the seismic recording instrument in the field is a significant step forward in maintaining consistency of instrument performance throughout a survey.

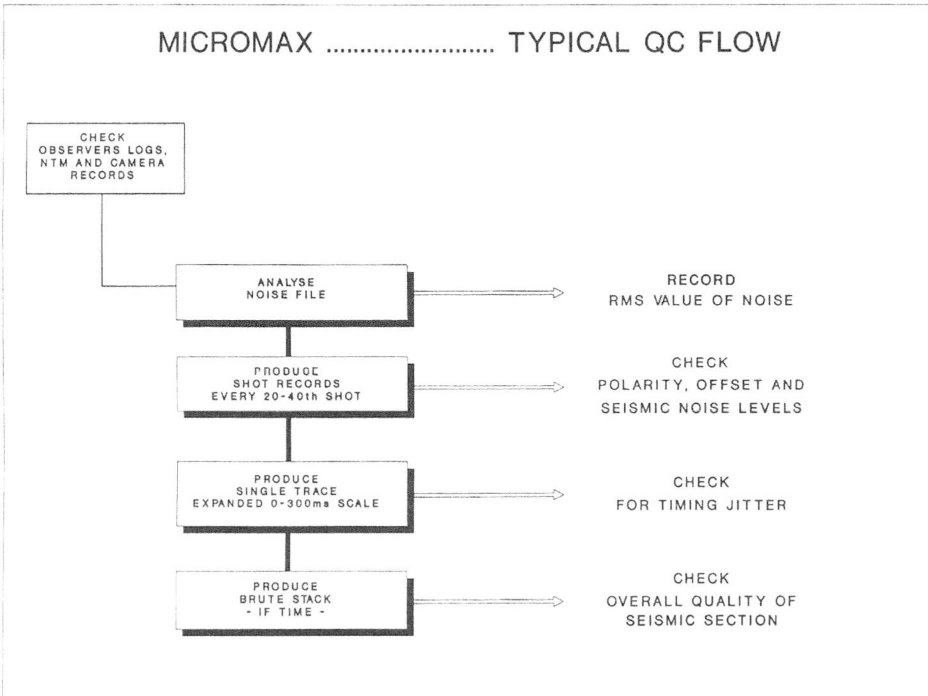

FIGURE 4

An example of a typical processing Flow for QC of the data. This Flow demonstrates a routine approach to QC analysis of each line (or part thereof).

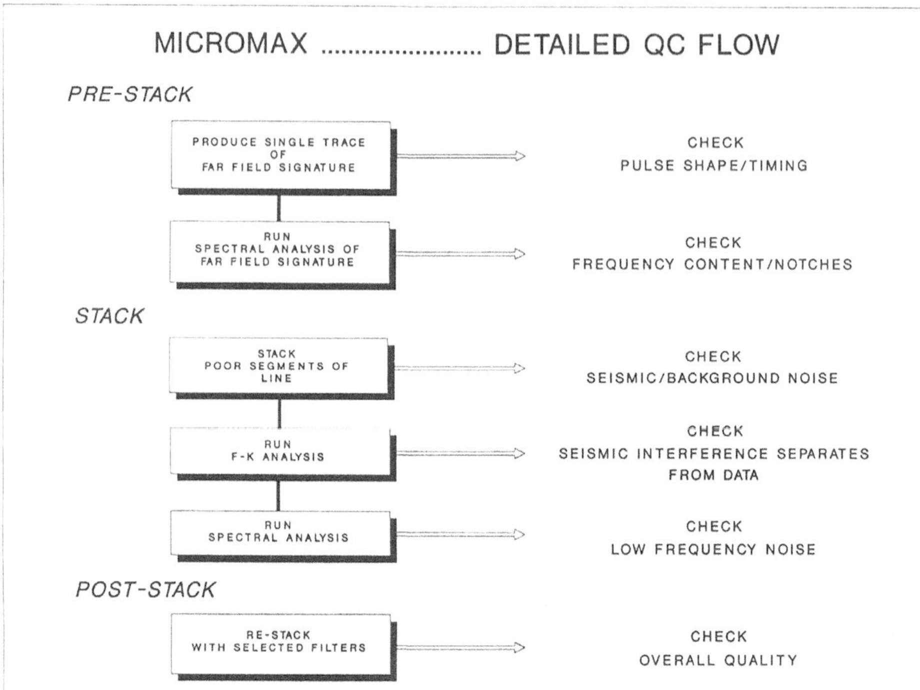

FIGURE 5

An example of a detailed processing Flow, where additional processing steps from pre-stack to post-stack are applied to the data.

The testing and selection of processing parameters at every stage are critical in order to judge the integrity of the SNR in the data.

Limitations during surveys are currently encountered in respect of the time and skill available to perform these Flows usefully and correctly.

POSITIONING ASPECTS .... KEY COMPARISONS

	EXPLORATION 3D	*'ENGINEERING 3D'*
STREAMER LENGTH	3000m	600m
IN-LINE CDP SPACING	12.5m	6.25m
TYPICAL TAILBUOY ACCURACY	+/- 15m	+/- 15m
TYPICAL COMPASS ANGULAR ACCURACY	0.5 degrees	0.5 degrees
MINIMUM X-LINE BIN SIZE	25m	5m?

ASSUMING

Typical Acquisition Parameters For 'Engineering 3D'

- Average Velocity • 1750 m/s
- Maximum Frequency • 200 Hz
- Maximum X-Line Dip • 20 degrees

THEN

A CROSS LINE SPACING OF 6m WOULD BE REQUIRED TO AVOID SPATIAL ALIASING

FIGURE 6 HYDROSEARCH

The traditional survey vessel positioning systems consist of various combinations of Radio, Acoustic, Satellite navigation information mixed with gyro and pitch/roll measurements which are used together in order to reference the survey platforms towing points.

Separate but similar techniques are used in Source/Streamer configurations to accurately determine the absolute and relative positions of individual Reflection Point Positions (RPP's).

Traditional 'Exploration 3D' survey techniques generally expect the RPP's to be known to a precision of approximately 25 metres.

Should 'Engineering 3D' surveys be adopted using these techniques for say field development directional drilling, then it is suggested a RPP's precision of approximately 5 metres would be required to achieve a useful 3D data set.

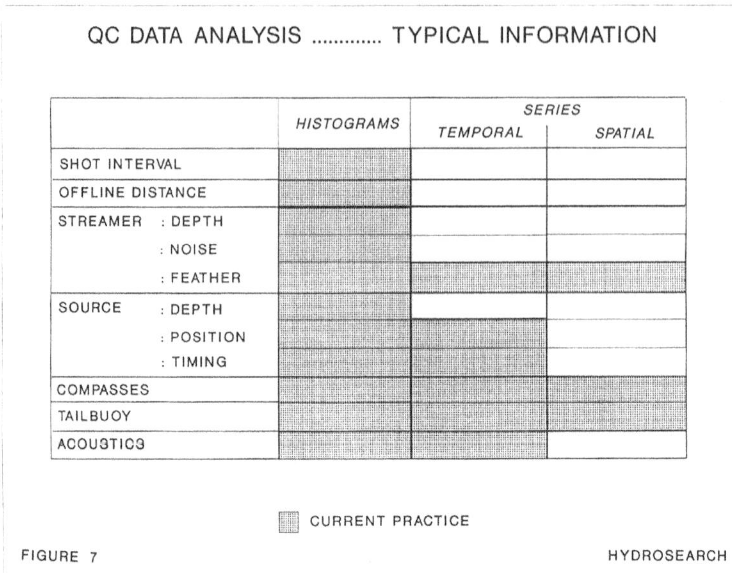

FIGURE 7

Large volumes of separate and varied data strings are generated and provided in different format for QC analysis. The typical examples shown in the figure are only a summary of this data strings.

Notwithstanding the time taken or skills required in the evaluation of this information, there are limitations for 3D surveys in Line-*by*-Line analyses using Mean/Average/S.D Histogram type information.

In a 3D data set, these Histograms only represent samples and do not, in isolation, demonstrate the Temporal and Spacial variations that can critically degrade a 3D data set.

By computing and sorting the Histogram data in a Line-*with*-Line approach for the purpose of quantifying trends and rates of change in critical parameters, then progressive and relative errors in the data set can be identified.

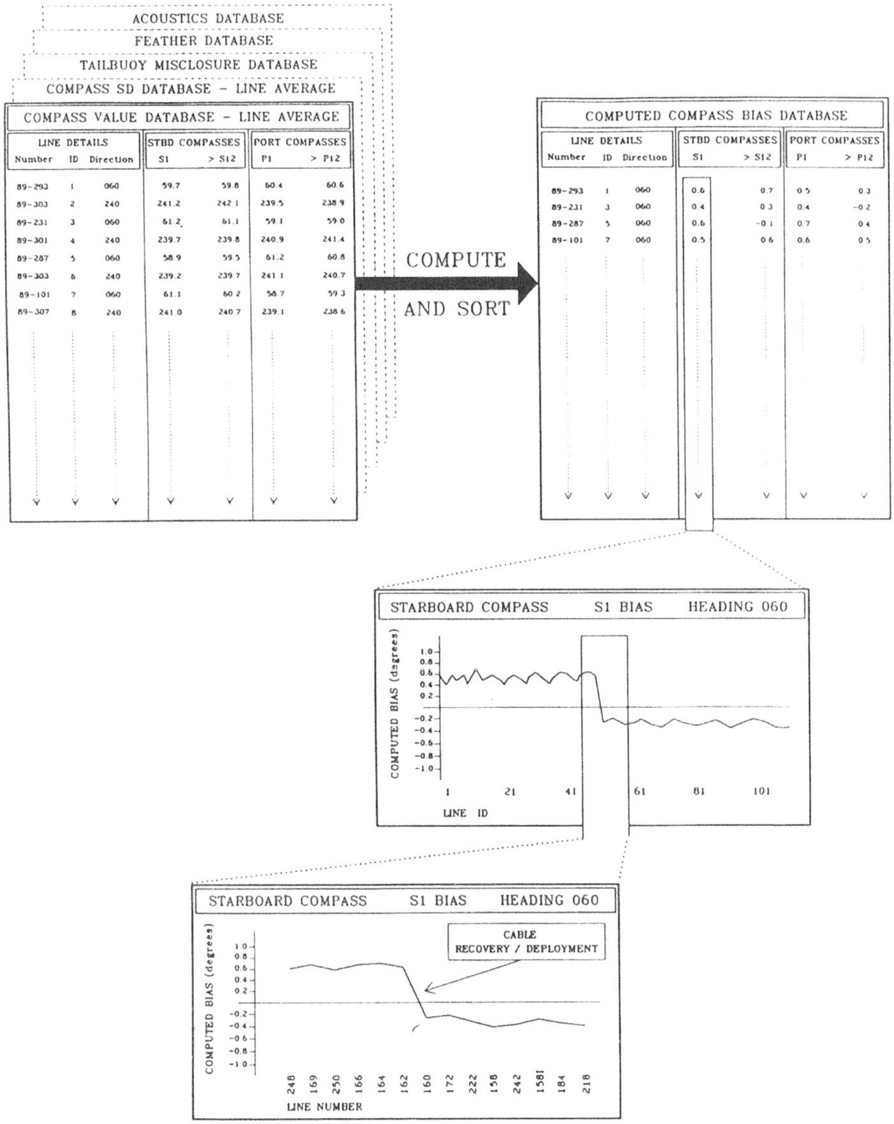

EXAMPLE OF TYPICAL ANALYSIS FLOW FOR 1 DATABASE PARAMETER

FIGURE 8 HYDROSEARCH

Demonstrates many of the QC analysis Flows that can be achieved when data strings are managed in Database form.

This example is simulated for an 'Exploration 3D' survey using dual streamers with 12 compasses per streamer.

The amount of data available for use in QC analyses can be enormous and current practice is to analyse data in this way onshore.

The capability to carry out these types of QC analyses as a routine offshore, is suggested to be rapidly approaching.

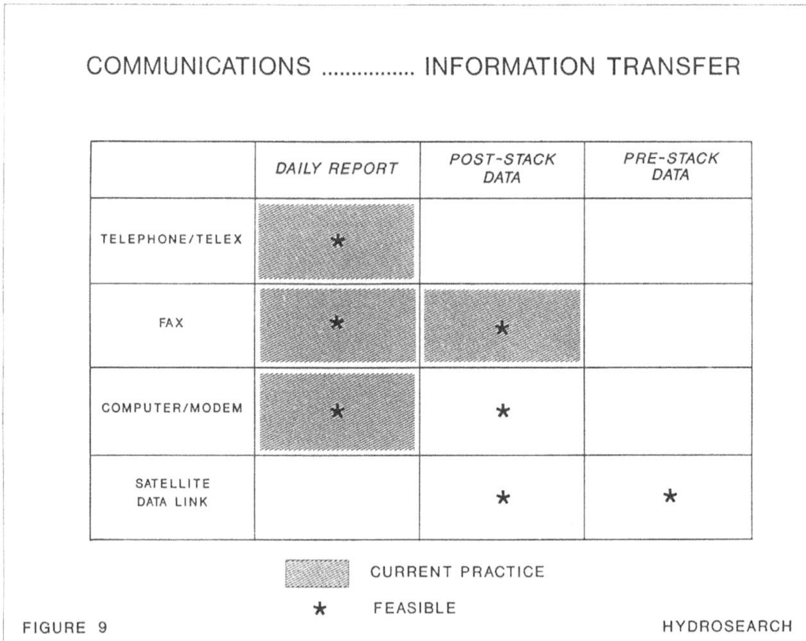

FIGURE 9

The current varieties of communications from vessel to shore are broadly sufficient to transfer information upto and including post-stack faxed sections.

The potential to transmit post-stack, or even pre-stack data by modem or satellite data link are currently available but are inhibited partially by technical issues but mostly financial. It is anticipated these inhibitions will diminish as satellite costs reduce.

It is suggested that if reliable and cost effective transmission to shore of pre-stack data (and QC analysis) are achieved as a matter of routine, then the current and projected difficulties in time/skill shortages for onboard processing may be alleviated by utilising comprehensive processing facilities onshore.

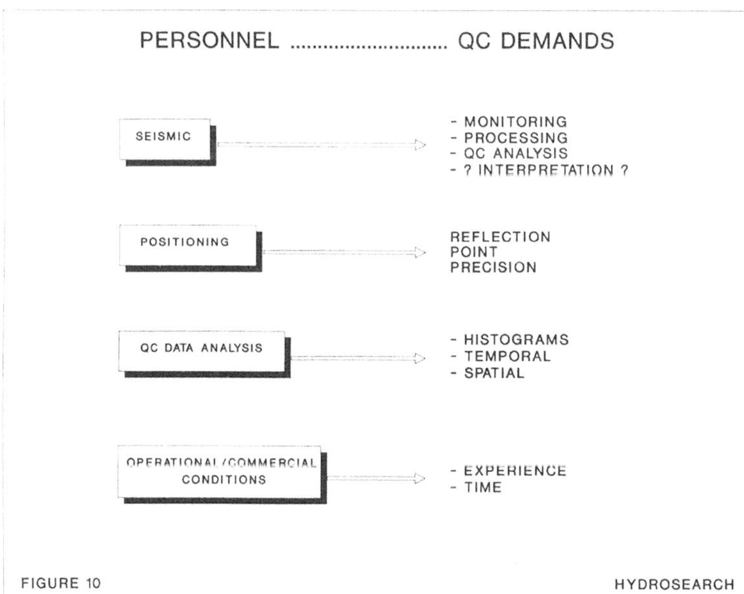

FIGURE 10

The demands on QC supervisors for 'Engineering' Seismic Surveys are very high and increasing. These demands are summarised in the figure.

The number of individuals who are fully experienced to QC supervise all aspects of seismic surveys is few. By sharing the demands between 2 or more personnel then greater attention can be paid to each technical activity.

However, the number of skilled personnel with the relevant experience are less than the current industries requirements in almost all offshore technical areas. Moreover, the immediate future appears bleak for recruitment as the number of new graduates are due to diminish over the next couple of years.

Skilled personnel are the key to the future success of Engineering Seismic Surveys, particularly 3D, for Shallow Gas determination.

Part III
Processing, Analysis and Interpretation

PROCESSING PROCEDURES FOR HIGH RESOLUTION SEISMIC DATA

K.P. Games

Gardline Surveys, Admiralty Road,
Great Yarmouth, Norfolk. NR30 3NG

ABSTRACT

The purpose of this talk is to highlight essential procedures for production of seismic sections for site survey interpretation. Other papers at this conference will concentrate on the use of individual techniques such as attribute analysis etc.. The main problem with high resolution gas hazard surveys is the very short turnaround time, and therefore it is essential that the processing procedures are correct and that no spurious effects are introduced.

It is particularly important to produce accurate sections if the subsequent use of workstations is envisaged during the interpretation stage. The interpreter must know and understand what has been done to the data he is working on.

Volume 25: Safety in Offshore Drilling, 103–131.

A brief discussion of acquisition of the data will be presented from the processing point of view. Emphasis will be on the provision of information to the processing house which can save a lot of time and potential errors.

Also of crucial importance is the role of those responsible for onboard QC who, if they do not understand the principles of seismic acquisition, may not make the correct decision when circumstances out of the ordinary arise.

An account of some essential QC techniques will be presented, along with examples of correct and incorrect ways of applying some of these procedures.

Finally, these topics are clearly illustrated using a typical example of a "bright spot", and some useful characteristics and information one can derive if the time and resources are made available.

INTRODUCTION

During the last conference held by SUT on shallow gas in 1987 I gave a presentation on Interpretation of Shallow Gas, and I ended that talk with a list of suggestions for the future. I thought it might be interesting to see if things have changed since then. The suggestions were :

1. ENCOURAGE R&D INTO SHALLOW GAS DETECTION TECHNIQUES
2. ENCOURAGE FEEDBACK FROM OIL COMPANIES AFTER DRILLING
3. STANDARDISATION OF REPORTING OF GAS HAZARDS BY SURVEY COMPANIES
4. REVIEW OF MINIMUM REQUIREMENTS FOR GAS HAZARD SURVEYS
5. DETAILED ANALYSIS OF ALL SITES WHERE GAS HAS BEEN FOUND

Taking each of these points in turn: 1) a lot of useful work has been going on particularly using workstations; 2) no real sign of change here; 3) no change; 4) it good to see that the UKOOA recommendations will be reviewed here tomorrow; 5) again, no apparent change.

Perhaps it would be good to set in motion some way of following through the ideas and benefits of these meetings?

AN OVERVIEW

This talk is about bulk processing of high resolution seismic data for site surveys. I do not intend to present any 'new techniques' - there are several other topics on the programme devoted to specific processes and problem-related techniques. Similarly, there are plenty of textbooks which describe the basic principles of processing, and for those who wish to delve into the physics of data processing I have included a reference list at the end of the presentation. My main objective in this short time available is to highlight some very important aspects of processing which are of particular relevance to site survey data and to the production of 'routine' quality sections. We are all aware of the difficulty introduced by the very short turnaround times in this part of the oil business. Great emphasis is often, rightly, laid on the specifications for equipment, vessel, interpretation, and reporting of the results. Oil companies usually put an independent observer on the vessel to ensure that data of a satisfactory 'quality' is acquired. But then what often happens to the seismic data is that it is processed with minimal input or control by the survey company or the client, and accepted for interpretation with very little knowledge of exactly what has been done to the data. I have been amazed over the years talking to many in the industry who clearly understand little about processing, yet who have an important role to play in either overseeing or even interpreting this data. I am not in any way being critical of processing houses in this - I know only too well the very difficult environment produced by the pressures to get data processed rapidly. Over the last 8 years I have had dealings with more than ten different processing houses, and have learnt a lot from many of their dedicated and enthusiastic personnel. But, in a commercial environment, in a process that is very personnel-dependent (as opposed to deep seismic processing which is quite machine intensive) I think that it is important to stress some basic ground rules which I know from experience have often not been followed.

Even given all the time in the world, no two processing houses will produce identical sections from the same data. Figure 1 shows a comparison of a test section processed by four centres (none of which are currently processing site survey data). In this example,

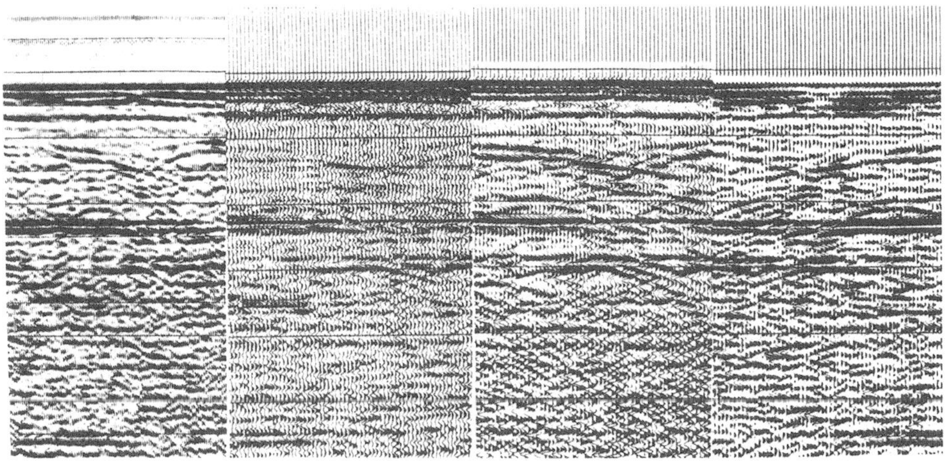

Figure 1. A sample of data processed by four different processing houses. The section on the left is at a different horizontal scale to the others, and the one on the right is displayed in the opposite direction.

I'm not trying to say one is necessarily better than another - it simply serves to emphasize that if such differences occur when there are no time constraints applied, how different can we expect data to be under normal conditions.

SOME THOUGHTS ON ACQUISITION

Much has been (and will be) said about acquisition at this meeting. My concern here is, given that a crew is out at sea shooting data under the terms of a contract, how can the processing house be best served? Clearly once the data has been acquired, that is what has to be dealt with onshore. So how can awareness by all involved help to give us better sections?

Of greatest importance is the careful and accurate recording of observers logs. Some things can be corrected at the processing stage if they are known - for example, missed shots, reverse polarity traces, static shifts; all can be sorted out, but the processor's life is made easier if he is aware of the problem, and it also avoids the possibility that such problems might not be picked up if the processing house does not carry out basic QC checks.

Then there is the problem of noise. No matter how rigid oil company specifications are, when a rig is waiting to move onto location the temptation is there to say 'shoot at all costs'. What does happen if the data is acquired in, say, a force 7-8? Well, apart from danger to personnel and equipment, Figure 2 shows just what it does to the data. This illustration compares a 'normal' direct arrival and seabed for a stable source, and what happens when the source is being tossed about in a rough sea. You will see that there are parts of the section which have much reduced power, which will clearly lead to problems in interpreting relative amplitude variations in a given reflector. And on top of that there is the noise on the streamer - seen in Figure 3 which shows some shots from the same survey.

Another thing to be aware of during acquisition is the importance of the correct source to near trace offset (see Figure 6). Ideally, this should always be kept to a minimal value - zero

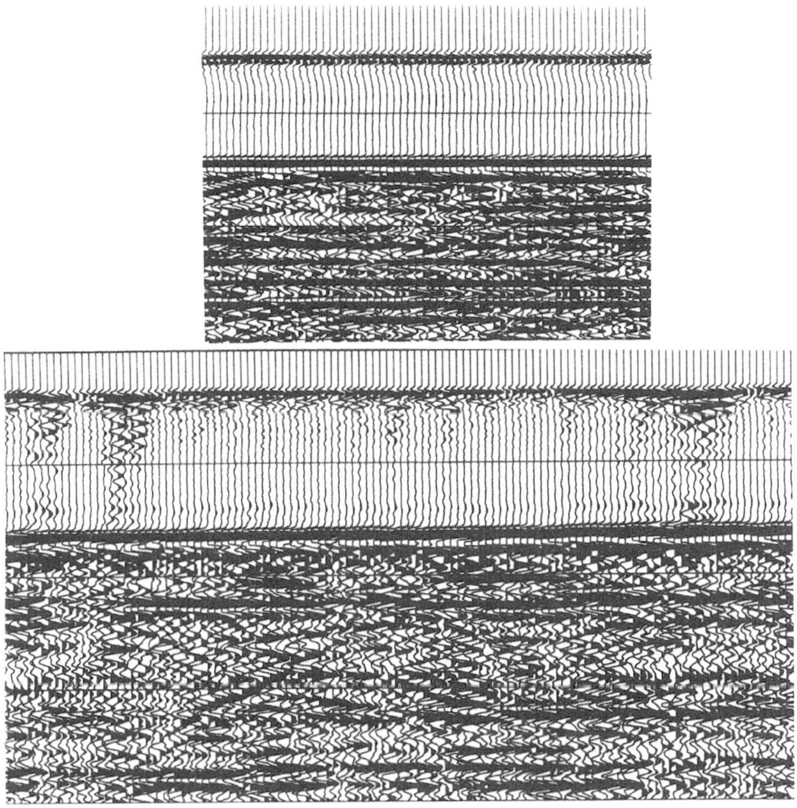

Figure 2. A 'normal' direct arrival and seabed return (top) compared with one from the same source but shot in a force 7-8. Notice the variation in source output and the corresponding gaps in reflectors where source power is greatly reduced.

offset is quite acceptable. But due to the noise (both ship noise and that caused by the towing) on the front end of the streamer, we do often have to get the streamer as far back as possible. If this is the case, the golden rule is that offset must be less than half of the water depth - otherwise the angle of incidence will cause severe degradation of the near-seabed data.

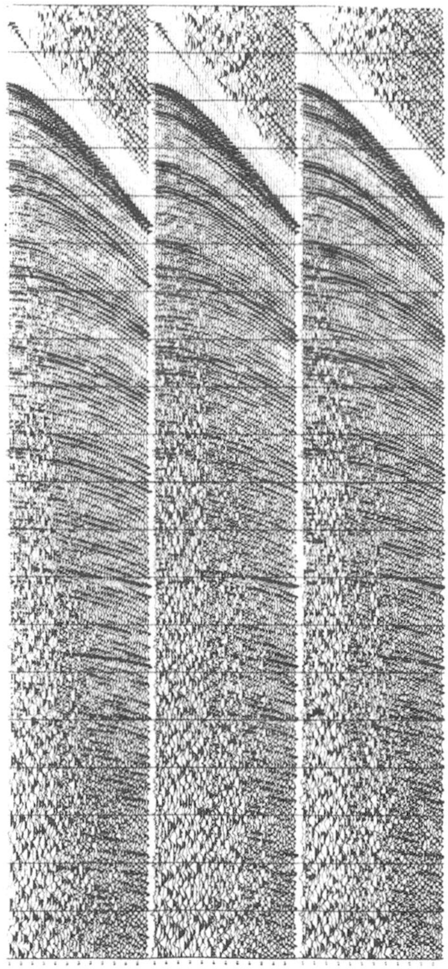

Figure 3. Three shot records from the same data set shown in Figure 2. Note how noisy the near 18 traces are, particularly below 500 ms.

A final point to remember is the frequency requirements of the survey and their consequent implications. So for example it is no use recording 0.5ms data if you use a 3m tow depth because of the notch produced by the 'ghost', and also the source and streamer tow depth should be the same (since the ghost affect applies to both).

AT THE PROCESSING HOUSE

As I have already suggested, I think the biggest problem facing site survey processing is how to produce the data in the time required without jeopardising quality. Some basic QC checks are therefore required:

Near Trace Plots (Figure 4)

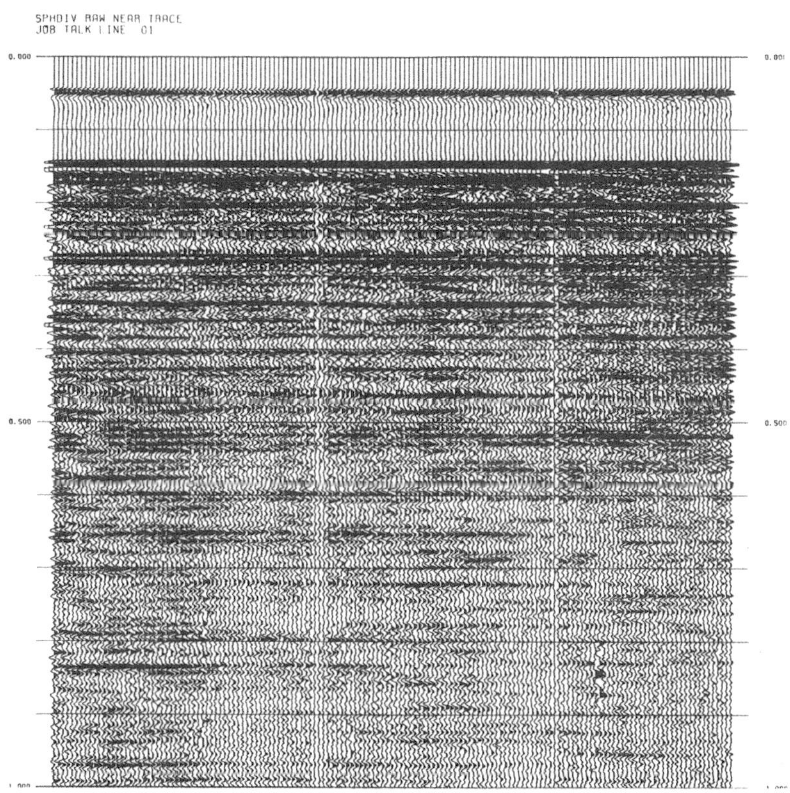

Figure 4. An example of a Near Trace display. This can be used to check offsets and to identify missed shots and noise problems.

These serve firstly as a check that offsets are correct - data is gathered at an early stage in processing, and if the geometry is not correct not only will the data be degraded, but having to correct this is costly. Other things to be gained are a look at the source stability, timing problems, missed shots and a first look at the geology of the site. It will also give an idea of what sort of testing for gain correction may be needed.

Shot Displays (Figure 5)

These help to show the status of the streamer (noise, dead or intermittent traces, sea state, tow depth variations). They will also show up potential problems with direct arrival, and will suggest an initial mute which can be used in the preliminary stages of processing.

The price for getting such basic checks wrong can be horrendous. No matter how well observers logs are filled in, it is imperative that checks on offsets etc. are made. I can recall one incident where there was a combination of a static shift (not recorded on logs), a precursor, a direct arrival coinciding with the seabed, and a sandwave field over the site! This was a very complex site to sort out, but because very poor QC checks were made at the start, the data was presented with the seabed appearing just above the first multiple, and it literally took months to present the final sections.

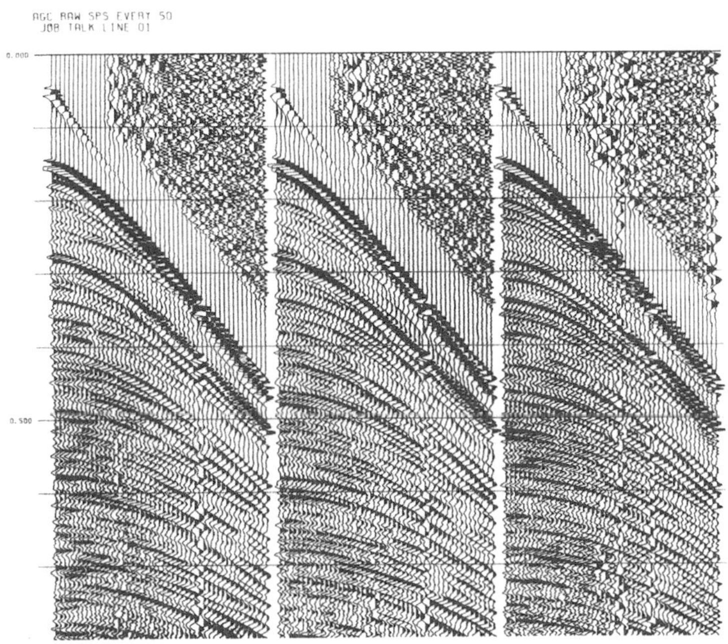

Figure 5. Three shots displayed to act as a monitor for the streamer status (such as trace 33 here which has reverse polarity) and to give an initial idea of the mute.

Brute (or Raw) Stacks (Figure 6)

Brute stack sections, produced usually with minimal processing and a single velocity function, are a very important stage in processing. Obviously they are a very good QC check, providing us with a first look at the stacked section, and highlighting potential problem areas. But they are more than that. We all know that data can be 'invented' by over-processing. It is important to see the data produced at this stage and compare it eventually

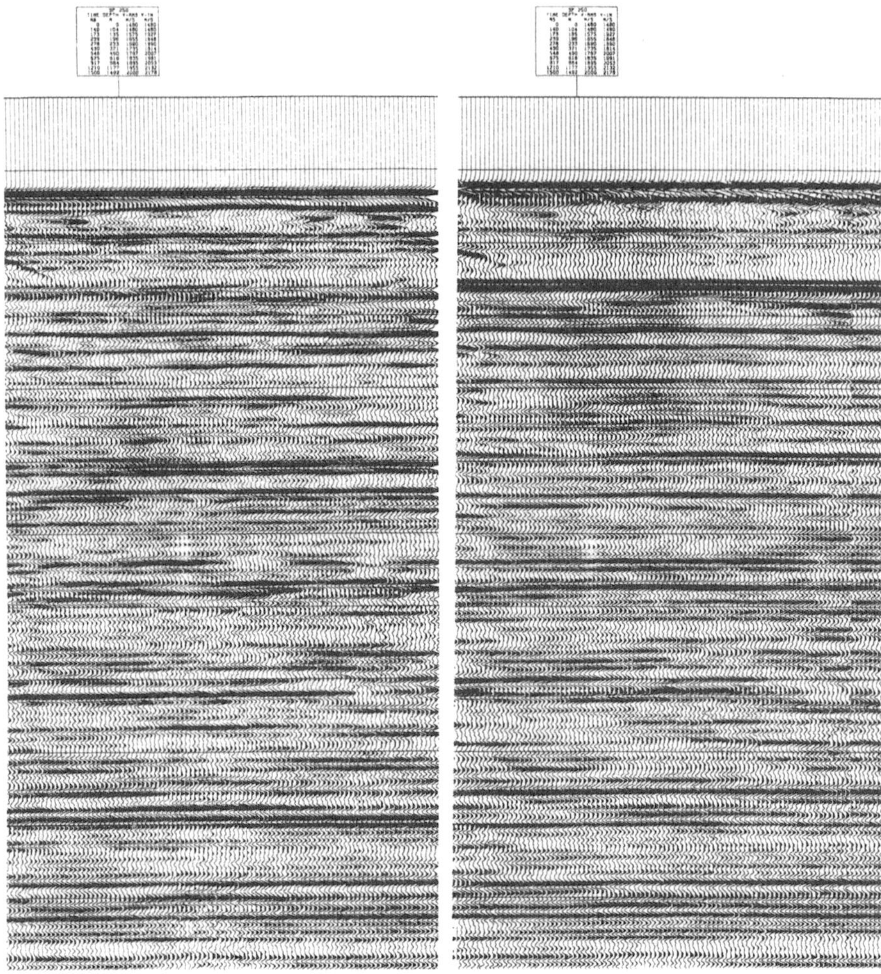

Figure 6. A brute stack produced with a single velocity function and gain correction only. The example on the RHS is identical to the correct brute on the left other than the fact that a wrong offset value was used to gather the data. Note how the shallow data is degraded and how strongly the multiple stacks in.

with finally processed data. I have seen spurious effects introduced by processing - the classic example is the effect of badly picked velocities, but other dangers can be incorrect DBS and DAS, FK processes, and migration. In the example shown, the same section has been reproduced using identical processing apart from using a wrong value for the offset. Notice in particular the way the multiple stacks in, which would certainly cause problems in the later stages of processing.

But it is also of great help to 'see' the data and put velocity picking in context of the geology as shown on the section. Sometimes it is necessary, due to time constraints, to perform interpretation on brute stacks. Provided the brute velocity function is good (and there are no severe problems on the site due to, for example, multiples) this can be acceptable if the client is made aware of the potential limitations of the interpretation and if he is informed if the final sections show any change of emphasis in interpretation.

Preservation of Amplitude Information

The amplitude of a reflector depends on the acoustic impedance contrast at the reflecting interface. However, there are many other factors in the life of a seismic wave which also affect its amplitude - noise, multiples, absorption, variation of reflection coefficient with angle of incidence (uneven or dipping reflector), interference, energy scattering, etc.. The crucial thing in all processing, but of great importance to high resolution data, is not to add yet another variable to this list.

The basic correction **always** applied is for spherical divergence. Figures 7 & 8 show a RA display of raw stacked data and a display of the same data with spherical divergence only applied. Alongside each display is a plot of amplitude against time for the first four streamer channels.

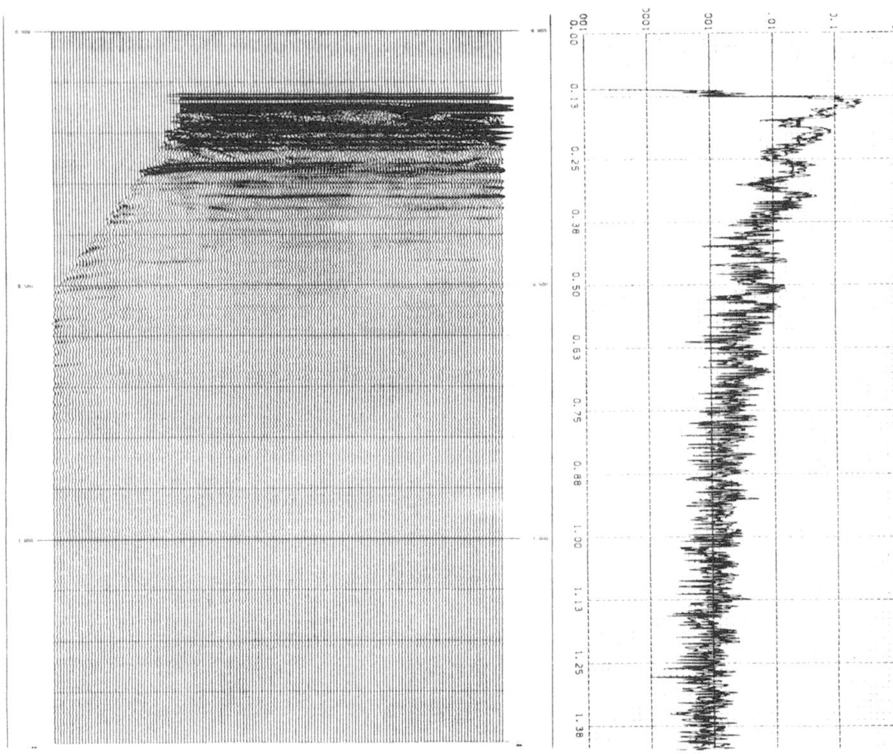

Figure 7. A relative amplitude display of raw stacked data with no gain correction, with a display of amplitude against time for channels 1-4.

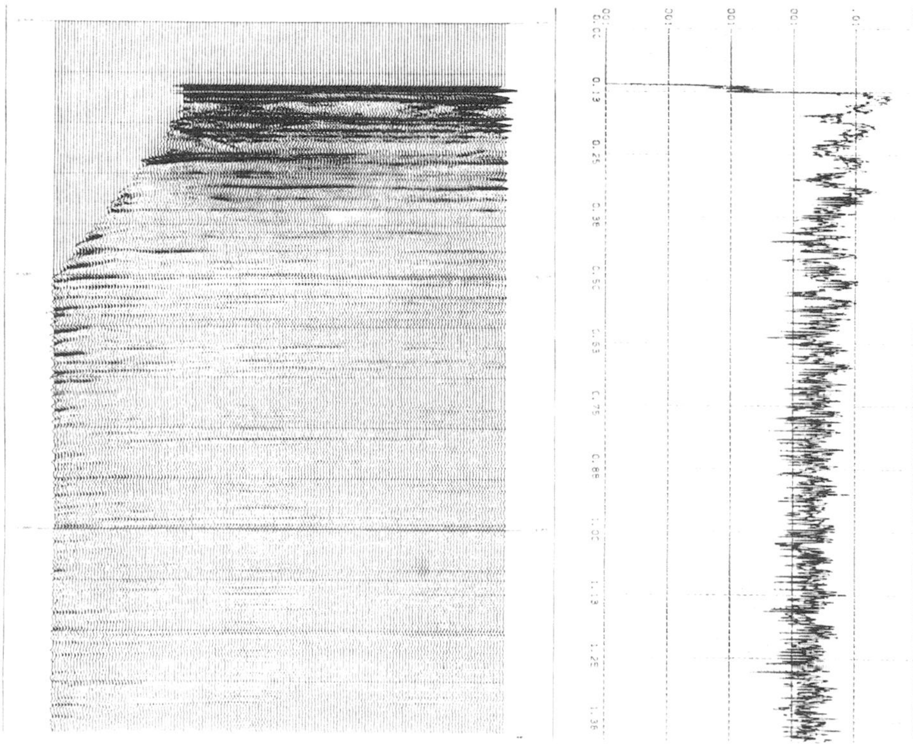

Figure 8. A relative amplitude display of stacked data with spherical divergence applied, with a display of amplitude against time shown on the right.

A problem specific to high resolution shallow data is that the seabed reflection amplitude is so much greater than the amplitude of events just below the seabed in which we are interested. So it is necessary to apply a careful gain correction (usually an exponential function) to make the range of amplitude values such that the display can cope with the objectives of the processing. Figure 9 shows how such an exponential function affects the previous examples.

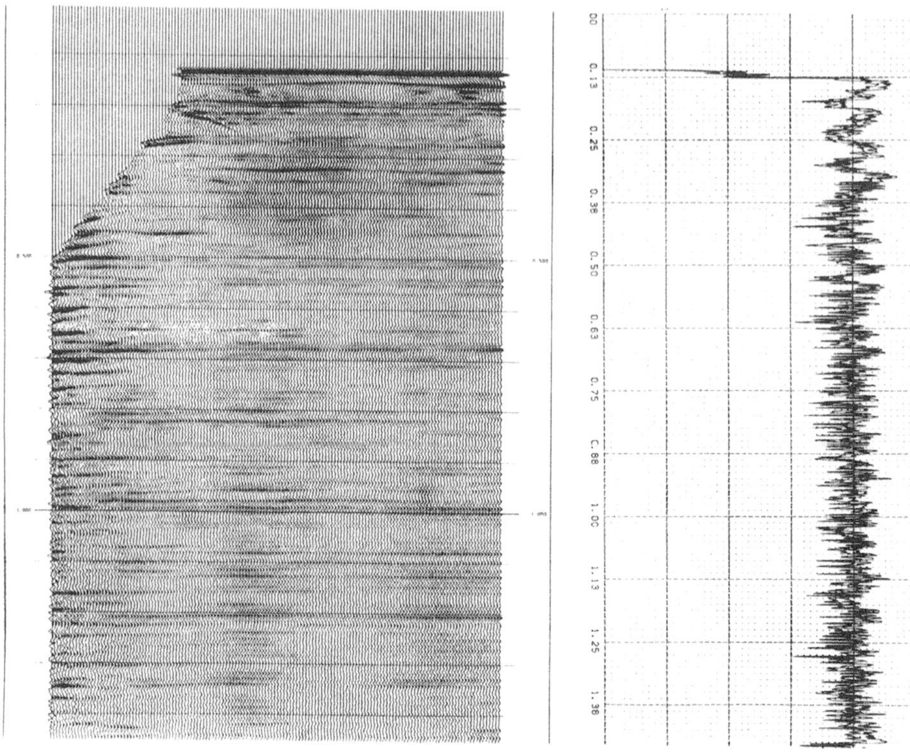

Figure 9. A relative amplitude display of stacked data with spherical divergence and an exponential correction applied, with a display of amplitude against time shown on the right.

Recent studies suggest that a correction for amplitude versus offset can be of importance for specific analysis of bright spots - AVO can identify the zones where the reflection strength changes as the angle of incidence changes.

Relative Amplitude Sections

This is a natural follow on from the last topic. RA displays are important in assessing site survey hazards - particularly the identification of shallow gas. Often changes in amplitude of a gas-charged sand can be quite subtle - for example there is little change in reflection coefficient between 5% and an 80% gas content. So it is vital that we do not introduce spurious effects in our RA displays. Exactly how a RA display should look is to some degree a matter of taste - some interpreters like to see more than others on these sections. Some have argued for a single display only, RA but with sufficient information preserved to be able to perform a full interpretation. However, most requirements are for both RA and Equalised sections, in which case the RA section can be used to highlight those zones of interest from a hazard point of view.

Figures 10 & 11 show what could be interpreted as a variation in amplitude in the data at around 260-350ms, but is in fact produced by an incorrect velocity function.

If these RA sections are correctly produced it is possible to identify such features as phase reversal, variations in amplitude, gas-water contacts etc. A good example of such features will be shown later.

A very useful way of displaying RA data which should always be considered if phase-reversed events are suspected, particularly if the source has a minimum phase output, is to reverse its polarity.

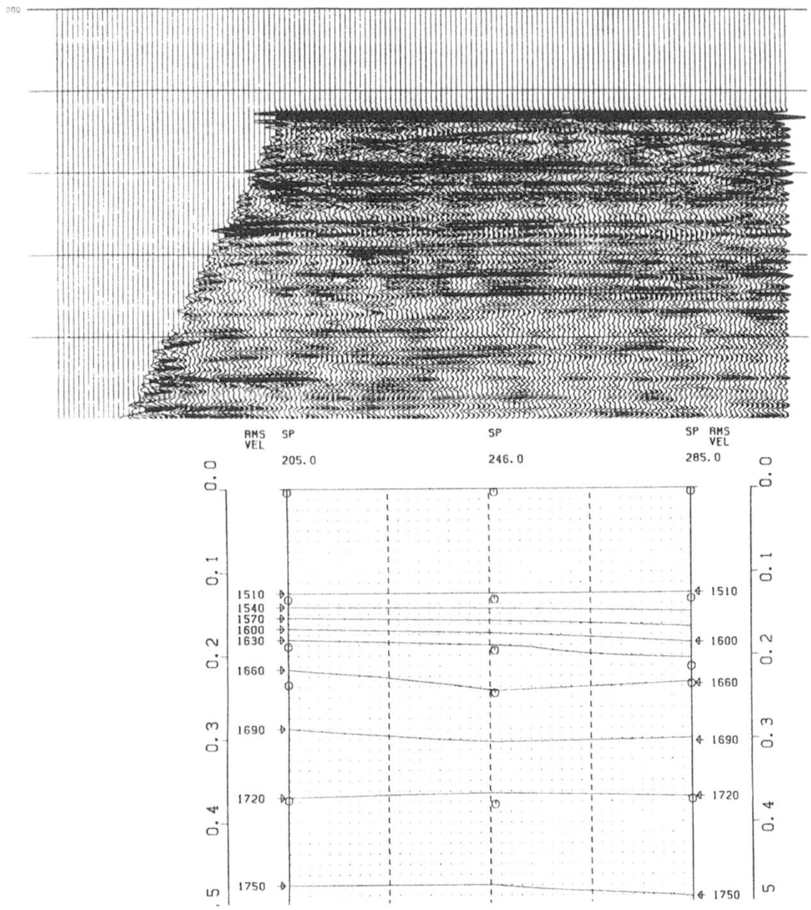

Figure 10. A relative amplitude display with, inset, a plot of an RMS velocity profile derived from the three velocity sites. Contour intervals are 30m/s.

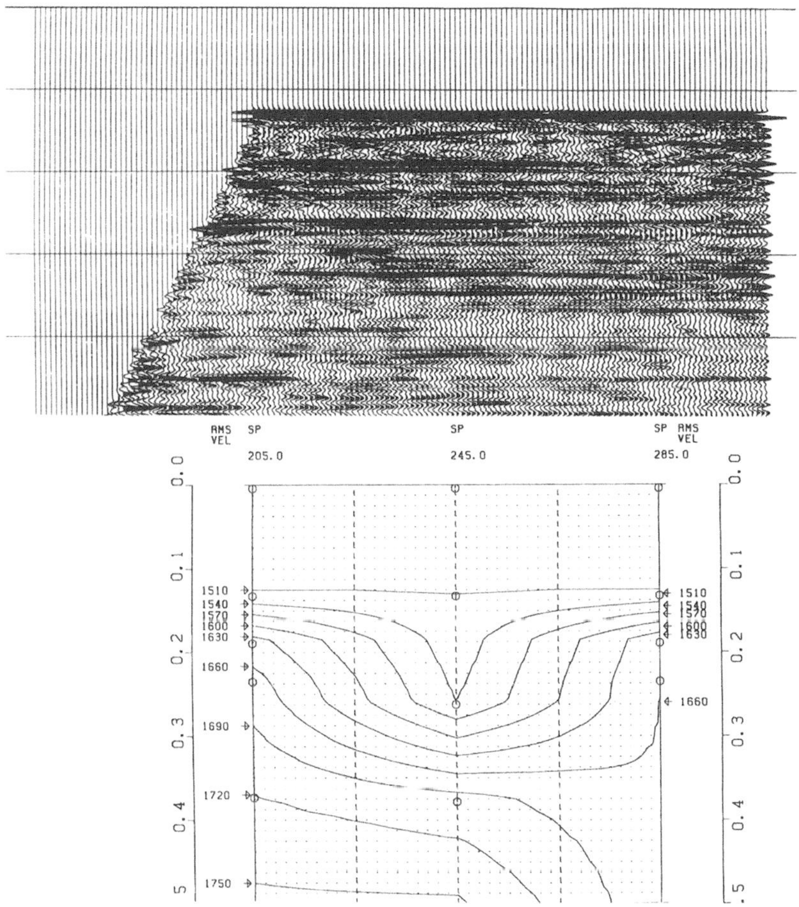

Figure 11. A relative amplitude display of the same section shown in Figure 10, but with an incorrect velocity function at the second velocity site. Notice the changes in the events between 260-350ms which could be wrongly interpreted from this section - the RMS velocity profile immediately shows up the bad velocity picks.

Velocity Analysis

The application of velocities to this high resolution data is probably the most important process in terms of what it does to the data. As Figure 11 showed, a wrong velocity can induce effects which may lead to a misinterpretation of the data. Such things do happen - far too frequently for comfort. I can recall one example where a 1.5 second record had a velocity site showing 4 picks-one on the seabed, one on the seabed multiple!, one at the end of the data (hence somewhat of a guess), leaving just one valid velocity pick throughout the whole record length. Another example was a site which had obviously been picked by two different people ; half of the lines had 6/7 picks per site, the other half had about 12 picks per site!

How should these velocities be carried out? Some companies are large enough to have a 'team' of velocity pickers. This is quite a luxury, as there is no doubt that velocity picking can be tedious, but it is important that whether they are done by such a group or by individual processors as and when necessary, all must have an awareness of the type of data they are working on and this process must not be done in isolation.

My personal opinion is that to perform accurate and reliable velocity analyses, the following minimum requirements should be adhered to :

1) At least two different functions should be used for example : Contoured velocity spectra
   Constant velocity stacks
   Corrected gather displays
2) Velocities must be picked in the context of the data set - hence the need for brute stack displays.
3) As much as possible should be known by the processors of the geology of the site - so such information, where available, as well data, deep seismic data, BGS (or equivalent) charts, should be consulted.

A very useful check to see that no spurious 'picks' are produced is to display an RMS velocity profile for the whole section. Figures 10 & 11 include such a profile for the first 500ms of data, and the wrong velocity pick is easily identified.

Other Processes

There is not enough time to discuss any other processes in any detail. Clearly the processor has available for use the whole range of seismic techniques with which we are all familiar. The most limiting factors which will often decide whether such things as migration or FK processes are to be used are time and cost. Many of these routines involve extensive testing which just becomes impractical in the time scale of the project as a whole. Also it is important to bear in mind the objectives of the survey. If these can be met by using a simple, effective processing sequence, then this is clearly the best approach to take. On the other hand, site survey data sets are becoming of increasing interest to the explorationists, and in these cases we may need to use a far more involved set of parameters to meet specific requirements.

Whatever the final sequence chosen, I feel it is important not to 'over-process' the data. No process should be applied 'automatically' without testing to see if a) it is needed and b) if it does indeed improve the data.

A CLASSICAL "BRIGHT SPOT"

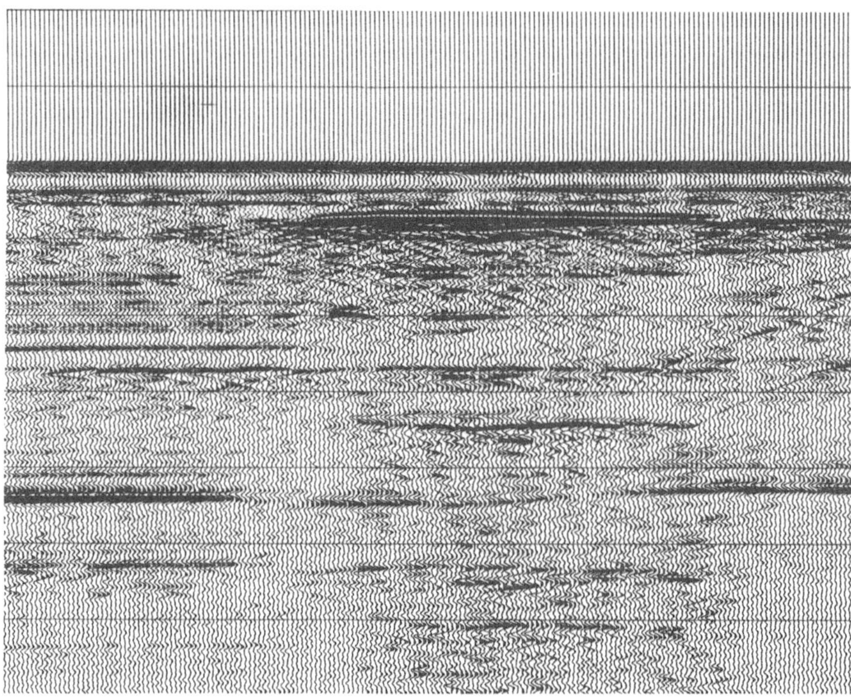

Figure 12. A brute stack, using a brute velocity function, of a classical "bright spot" exhibiting many of the characteristics of such an anomaly.

To end I would like to show a few examples of some processing of a very shallow bright spot, with particular emphasis in highlighting a few of the points made earlier in the presentation. Figure 12 is a brute stack RA display of the amplitude anomaly, which has had gain correction applied and has been processed with a single brute velocity function. This anomaly is so strong that it would be very hard to miss it even with poor processing. So Figure 13 shows the same brute stack produced using a slow velocity function - notice how strong the multiple appears, and how much of the shallow data apart from the bright spot itself has been lost. Figure 14 has been stacked with a wrong offset and a velocity which is too fast - again most of the shallow information is severely distorted.

Figure 13. The same section as Figure 12, but processed with velocities which are too slow - note the loss of detail and the multiple strength.

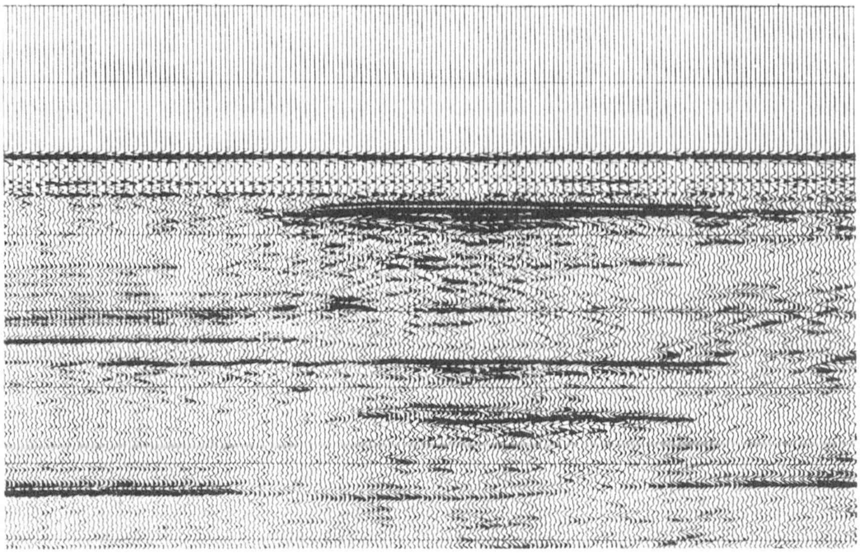

Figure 14. The same section as Figure 12, but to show the importance of getting the basic QC check right, this has been produced using a wrong value for the offset and a brute velocity function which is too fast. Most of the shallow data is severely distorted.

Such a strong anomaly will have a very low velocity field associated with it if it is indeed gas-charged. This can be identified as a velocity pull-down effect (Figure 12), but is also well shown in Figure 15. The top illustration shows a sequence of seven CDP's along the portion of data shown in Figure 12 with the brute velocity function applied to each one. The bottom display is the same CDP's with a picked velocity function on each. Note the event on CDP 2 at 350ms has been flattened by the application of the correct velocity, as has the bright spot itself on all CDP's.

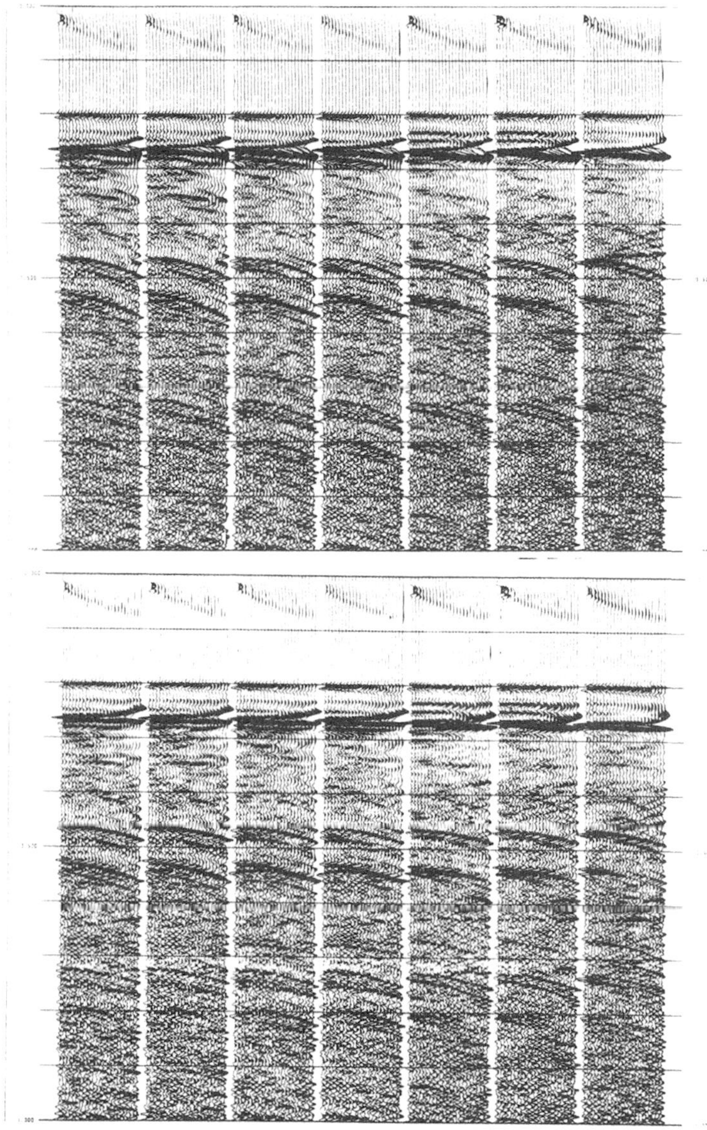

Figure 15. Seven CDP's from the data shown in Figure 12. The top set have been corrected using the brute velocity function, the bottom with a picked velocity on each CDP.

To show just how large an effect this change in velocity really is, the picked velocities have been displayed as an RMS velocity profile. Figure 16 shows such a profile for the rest of the line, excluding the 'bright spot'.

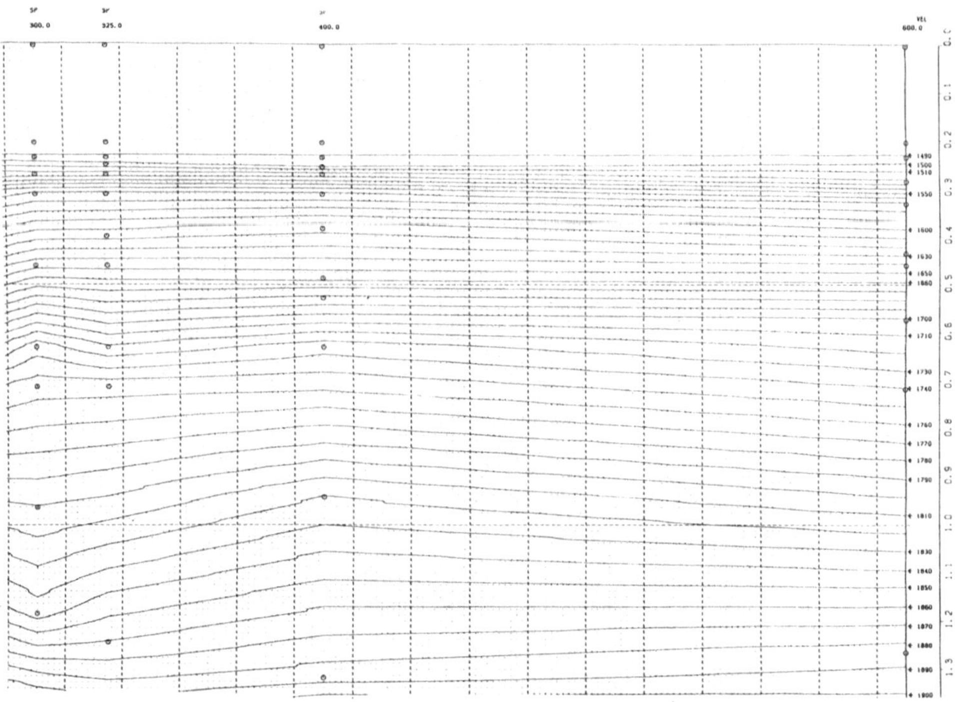

Figure 16. An RMS velocity profile for the whole of the seismic line excluding the bright spot.

Figure 17 shows the RMS profile over the bright spot itself. This is displayed at the same scale as the finally processed section (Figure 18) so that is can be superimposed on the data to show the shape and velocity effect of the anomaly. The contour values are in 10 m/s intervals and the anomaly itself shows as a 40m/s (RMS) inversion.

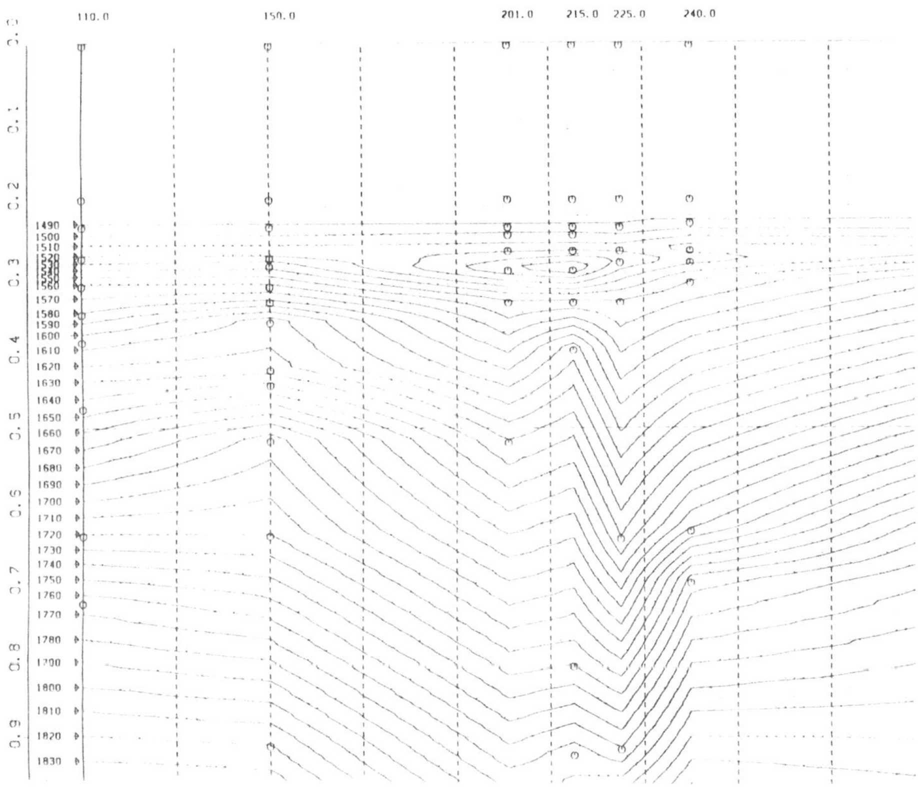

Figure 17. An RMS profile over the anomaly itself. Note that the contour intervals are 10m/s, and the velocity decrease due to the gas is 40m/s (RMS), which results in interval velocities of between 1100 and 1200m/s.

Finally, Figure 19 is an example of a Relative Amplitude displayed with reverse polarity, and it shows how the Bright Spot is clearly phase-reversed compared with the seabed return.

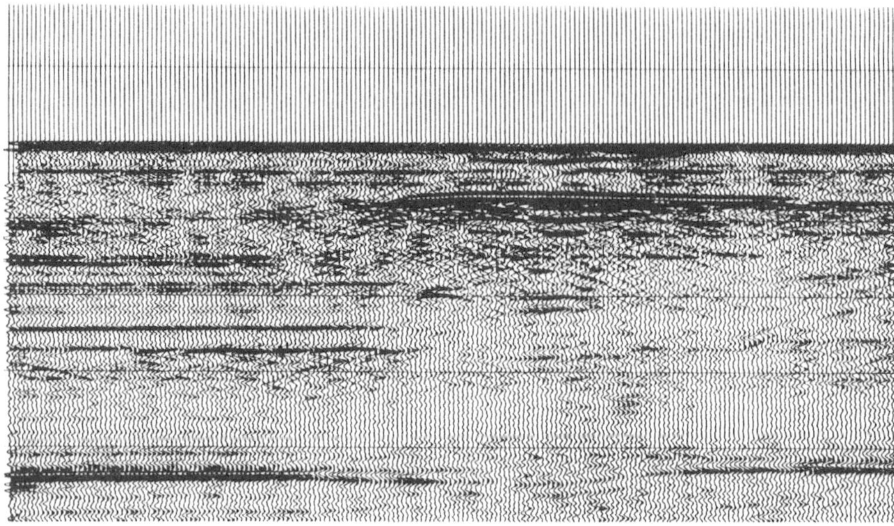

Figure 18. The final stack (relative amplitude) of the bright spot displayed at the same scale as Figure 17.

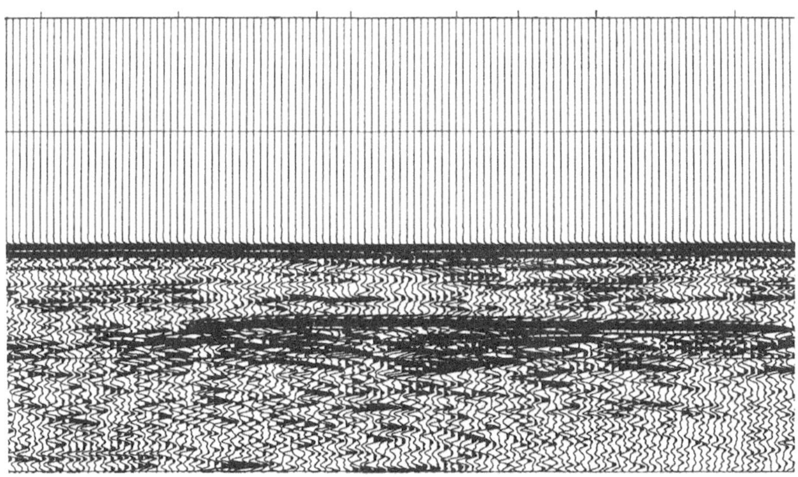

Figure 19. The same section as Figure 18, but displayed with reverse polarity, and showing the phase-reversed nature of the bright spot.

ACKNOWLEDGEMENTS

I would like to thank all of my colleagues for their support, and particularly Tony Bateson for the essential part he has played in setting up our in-house seismic processing. My thanks also to Gardline for allowing me the time to prepare this paper, and for the vision to make the considerable investment necessary to undertake such a major committment into seismic processing.

REFERENCES

R. McQuillan, M. Bacon, and W. Barclay Houston. An Introduction to Seismic Interpretation, Gulf Publishing Company.

R.E. Sheriff & L.P. Geldart. Exploration Seismology : Data Processing and Interpretation, Cambridge University Press.

A.R. Camina and G.J. Janacek. Mathematics for Seismic Data Processing and Interpretation, Graham and Trotman.

N.A. Anstey. Seismic Interpretation - The Physical Aspects, IHRDC.

SHALLOW GAS DETECTION USING AVO PROCESSING OF HIGH RESOLUTION SEISMIC DATA

A. R. Townsend and T. L. Armstrong

Total Oil Marine p.l.c., Crawpeel Road, Altens, Aberdeen AB9 2AG

ABSTRACT

At whatever location we proposed to drill, the shallow site survey invariably predicted the presence of potential shallow gas accumulations. Our credibility with the drillers was very low. We decided to investigate AVO (amplitude versus offset) phenomena in an attempt to discriminate between seismic "bright spots" caused by lithological contrasts and those due to gas-bearing sands. Case histories of three North Sea wells are presented with an analysis of how successful pre-drilling predictions have been. The deliberate search for AVO effects has had a major beneficial impact on our approach to shallow gas hazard detection. Extensive pre-stack analysis has revealed shortcomings in acquisition and processing, and pointed the way forward. We conclude that AVO analysis is potentially an effective tool in the reliable prediction of shallow gas hazards.

Volume 25: Safety in Offshore Drilling, 133–165.

INTRODUCTION

In our experience, there has been a tendency to interpret any high-amplitude seismic reflector as a potential "bright spot" and to infer the presence of shallow gas accumulations on this basis. Frequently, several shallow gas zones have been predicted at which appropriate drilling precautions were advised. A trend soon emerged that at whatever surface location we proposed to drill a well, the shallow site survey seemed to predict the likelihood of encountering shallow gas. Subsequent drilling invariably encountered no shallow gas zones and often, not even an increase in the background gas measured from drilling-mud returns. We were in danger of losing all credibility with our drillers.

We present case histories for three North Sea wells and analyse just how successful pre-drilling predictions have been. The results of these post-mortem analyses have had a dramatic impact on our approach to acquisition, processing and interpretation of shallow gas hazard seismic data.

WELL ONE

A high resolution digital seismic survey was carried out across the proposed well location. A final site survey report was prepared by the contractor and in it, we read the following:

> "A number of anomalous high amplitude events were noted on the relative amplitude sections, particularly around the proposed drilling location.
>
> These bright spots occur at four levels, 100m, 339-405m, 500-560m

and 800-880m below the seabed respectively. Elsewhere in the survey area occasional bright spots are found at these levels and at an intermediate level, 235-283 metres below seabed.

The uppermost bright spot level (100m) occurs within an acoustically transparent sequence (probably clay) and may represent a very strong lithological contrast (clay/gravel) or a shallow gas pocket.

The lower four bright spot levels are associated with continuous medium-high amplitude reflectors and are more likely to represent lithological contrasts, although due to their anomalous amplitude appropriate caution should be exercised."

In conclusion, our contractor stated:

"Shallow gas may be encountered at 100m, 339-405m, 500-560m and 800-880m and therefore precautions should be taken when drilling at these levels."

These conclusions predict a cumulative shallow gas hazard thickness of ca.220m, some 30% of the 26" drilling phase. As a direct result of these shallow gas predictions, a 12¼" pilot hole was drilled with a MWD (measurement while drilling) gamma ray/resistivity tool before the hole was opened out to 26" diameter.

No significant shallow gas was encountered.

The relative amplitude seismic section through the proposed well location is shown on Figure 1. The final well location was moved to meet

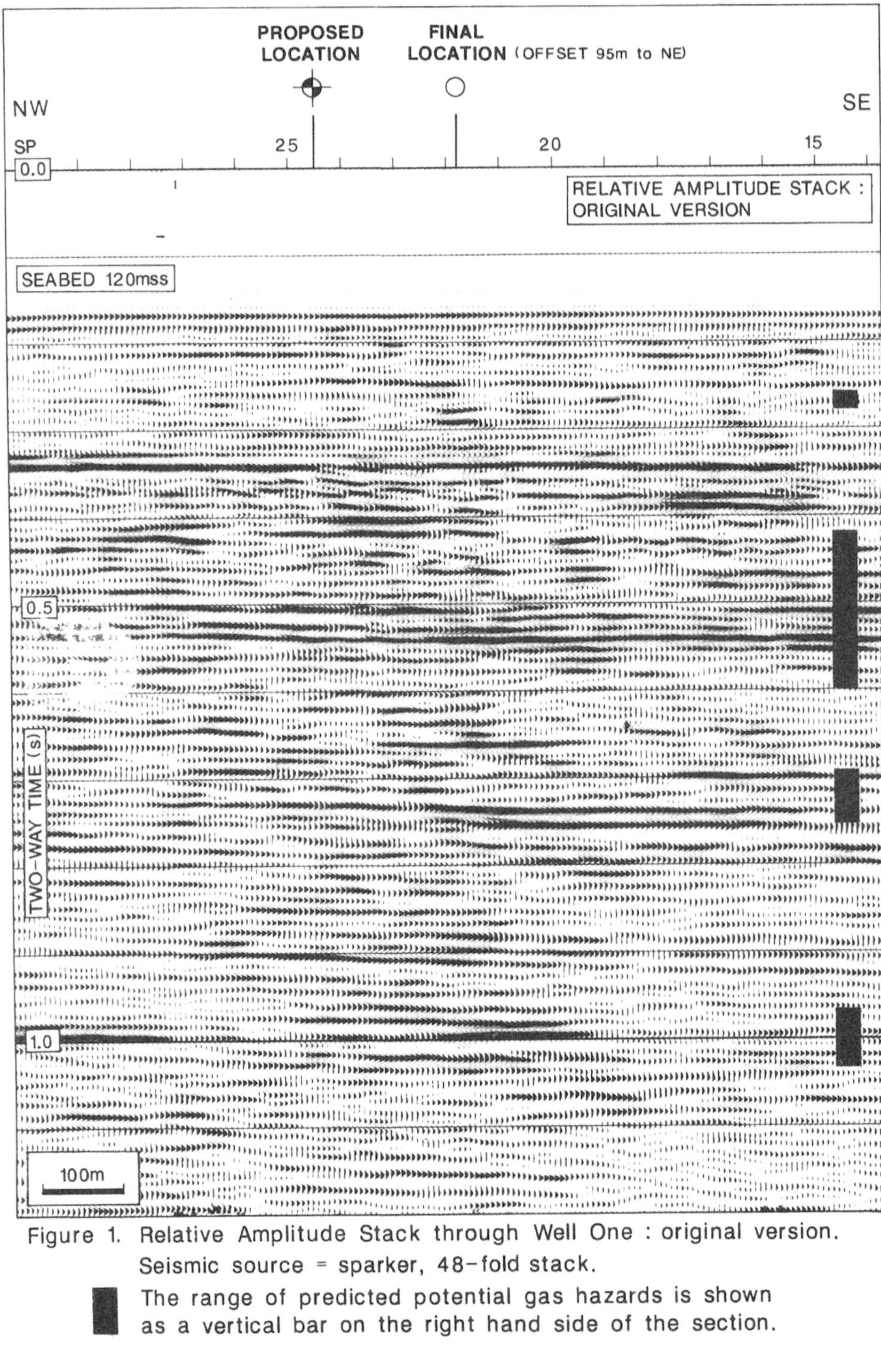

Figure 1. Relative Amplitude Stack through Well One : original version. Seismic source = sparker, 48-fold stack.

■ The range of predicted potential gas hazards is shown as a vertical bar on the right hand side of the section.

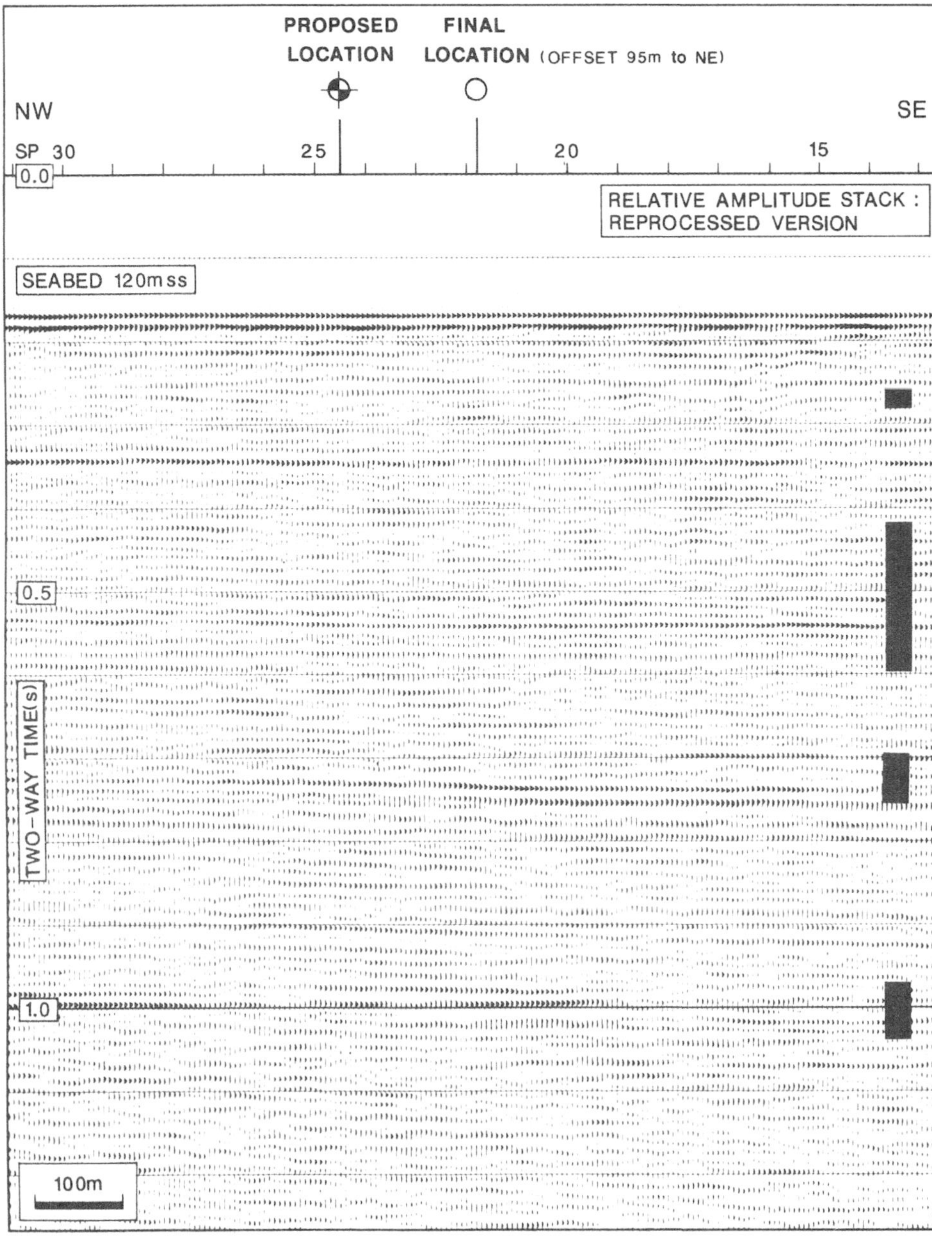

Figure 2. Relative Amplitude Stack through Well One: reprocessed version. Seismic source = sparker, 48-fold stack.

■ The range of predicted potential gas hazards is shown as a vertical bar on the right hand side of the section.

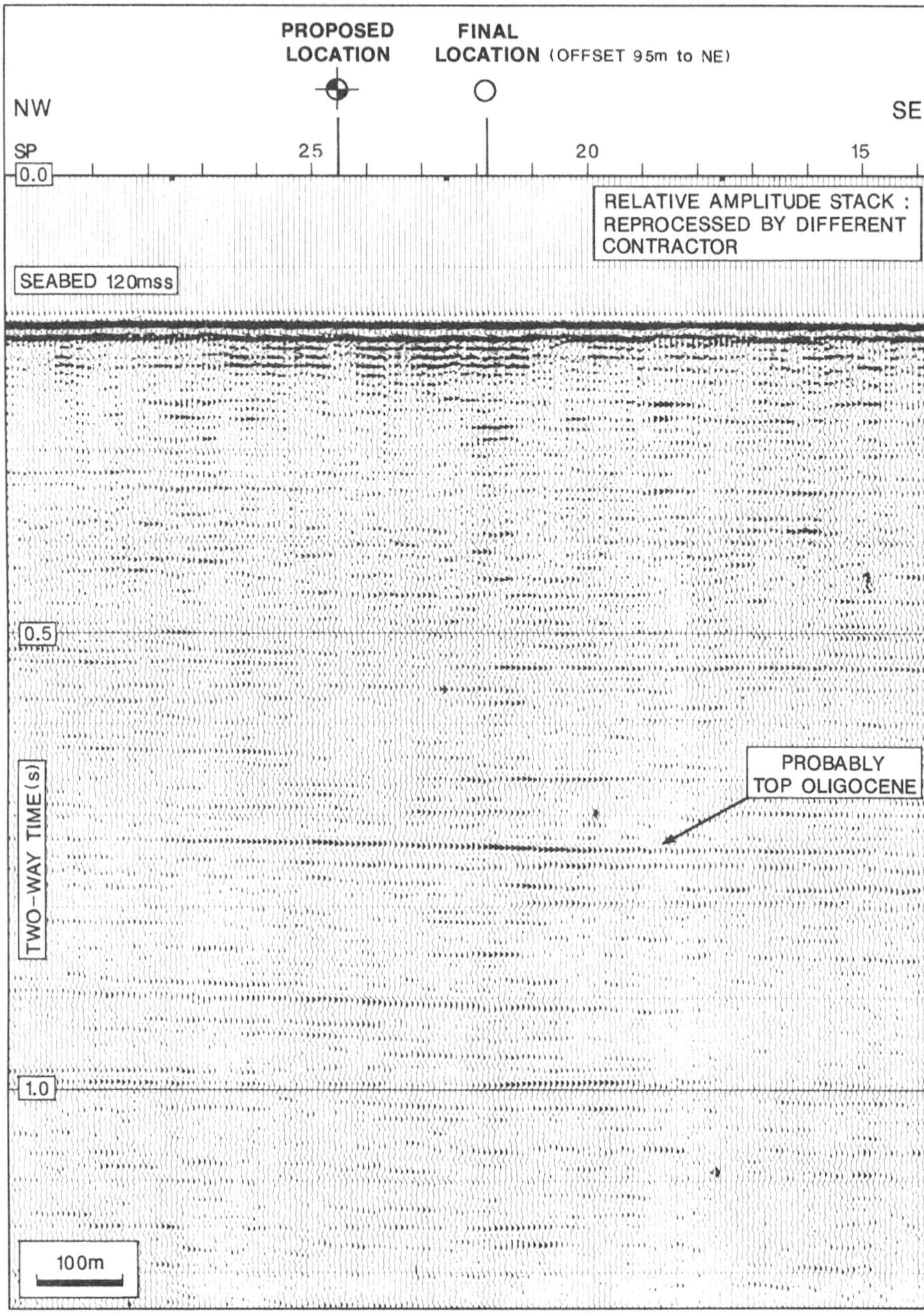

Figure 3. Relative Amplitude Stack through Well One : reprocessed version by different contractor. The Top Oligocene seismic reflector separates Miocene sands from underlying Oligocene clay.

exploration objectives and lies 95m to the NE of SP21.8. The contractor's report only discussed interpreted shallow gas hazards in terms of depth in metres below seabed. It was therefore impossible to relate predicted hazards to their respective "bright spots" without constructing a time-depth curve from stacking velocity data (note: we now insist that seismic times to interpreted shallow gas hazards are quoted along with predicted depths in metres subsea, mss). The seismic time zones equivalent to the contractor's predicted gas hazards are indicated on Figure 1. Despite no trace equalisation, the gain of the relative amplitude stack looks very high and clearly shows a number of high amplitude events, the so-called "bright spots", throughout the section. Attempts at amplitude versus offset (AVO) analysis proved inconclusive at this stage.

We were very unhappy with the appearance of the relative amplitude stack and wholly sceptical of the contractor's predictions of the possible occurrence of shallow gas accumulations in four discrete intervals below the proposed location. The seismic data were therefore reprocessed.

Figure 2 shows the new version of the relative amplitude stack reprocessed by the same processing contractor and produced some 40 days after the original version of Figure 1 (and, unfortunately, 32 days after spudding the well). The differences between the two sections are stunning: seabed multiples have been effectively attenuated, the overall gain level is vastly superior and the presence of bright spots has been dramatically reduced. Indeed, it can be argued there are no "bright spots" at all, entirely consistent with the absence of any significant shallow gas logged in the well. Figure 3 shows the same line reprocessed by another contractor.

The well did encounter a small gas peak of ca.1.0% total gas just below a depth of 966mBKB (seismic time equivalent estimated to be 990ms), the top of

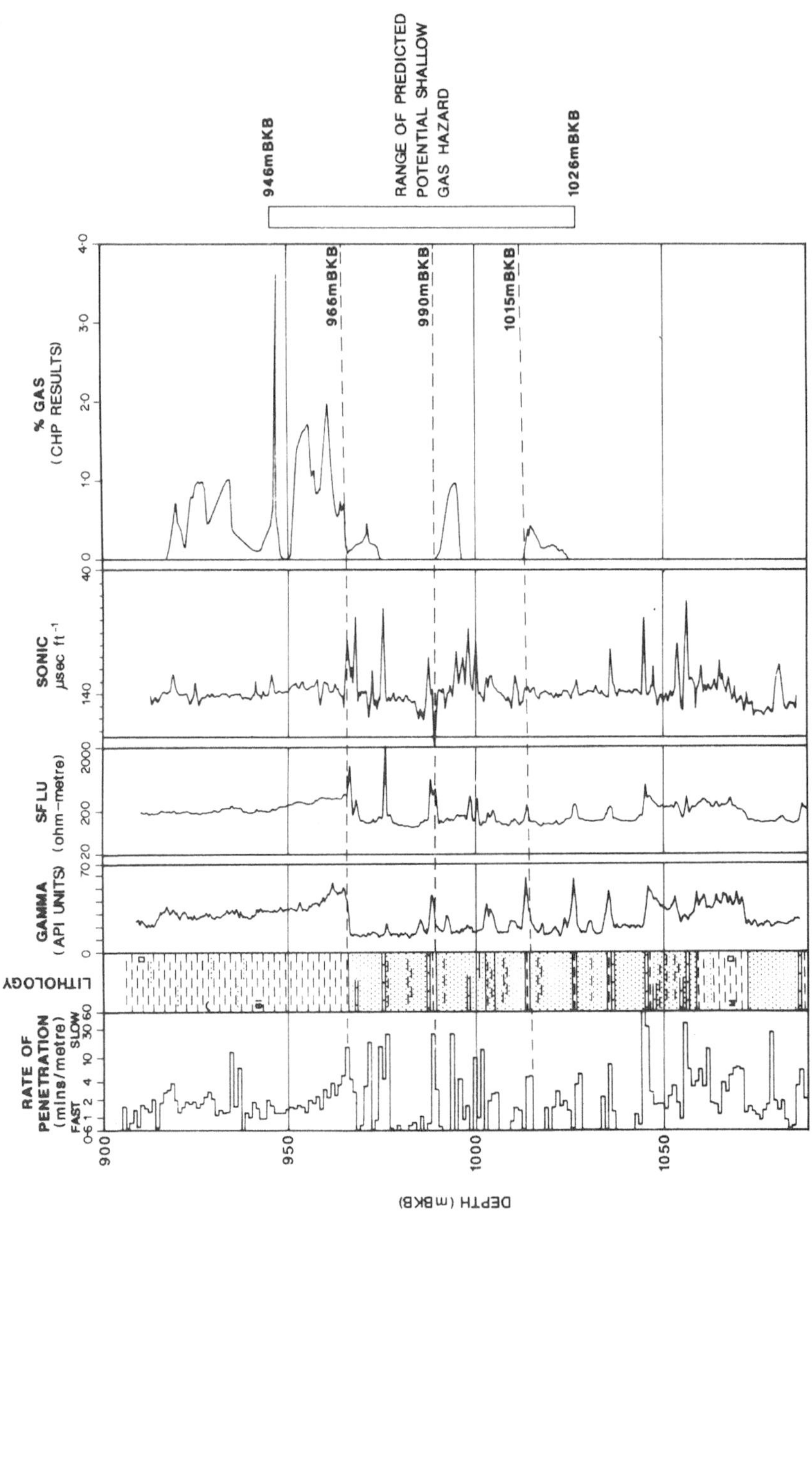

Figure 4. Well 1 : Shallow gas occurrences versus predictions.

a sand interval (tight with a calcareous cement in its uppermost 3m). The available logs for this section of the well are shown on Figure 4. The sonic log shows an increase in velocity from $2180ms^{-1}$ ($140\mu sec.ft^{-1}$) to $2470ms^{-1}$ ($124\mu sec.ft^{-1}$) and the high velocity interval is 3m thick. Using the formula of Widess (1973) for thin beds, we might expect reflection amplitudes in the range 0.77A to 1.08A for dominant seismic wavelet frequencies between 50Hz and 70Hz respectively (where A = the reflected amplitude from a single finite step in the sonic log). Since the density log (not shown on Figure 4) also shows an increase, we should expect to see a seismic response (positive reflection coefficient) from this thin velocity interval. This is observed on Figures 2 and 3. However, because of the small increase in total gas measured, we conclude that this interval is not a major shallow, gas-bearing sand.

Indeed, a close look at the logs of Figure 4 reveals that most of the measured total gas is associated with clay and claystone units. As drilling proceeds, mud filtrate invades the formations adjacent to the borehole and ahead of the drill bit, pushing the in situ fluids back into the formation until sufficient mudcake has built up to form a seal. The gas which finds its way into the mud comes from the volume of rock ground up by the drill bit, release of gas from the cuttings returning to the surface and by diffusion through the mud cake, even in the presence of a differential pressure (Desbrandes, 1985). In this case, the gas is most likely of biogenic origin.

The sand intervals have almost no gas associated with them. However, small gas peaks do occur immediately below mudstone seals at 990mBKB and 1015mBKB. The gas concentration measured for the sand unit below 1015mBKB shows a decreasing value with increasing depth.

CONTRACTOR ONE ORIGINAL SECTION (Figure 1)	CONTRACTOR ONE REPROCESSED SECTION (Figure 2)	CONTRACTOR TWO REPROCESSED SECTION (Figure 3)
1. DEMUX	1. DEMUX	1. DEMUX, 1ms SAMPLE RATE
2. GAIN RECOVERY USING INVERSE AMPLITUDE DECAY CURVES	2. GAIN RECOVERY USING FORMULA GAIN (dB) = $t + 1.5 * 20 \log_{10} t - 45$	2. AMPLITUDE RECOVERY 25dB/sec TO 1.0s
3. FK FILTER	3. FK FILTER	3. FK FILTER
4. 48-FOLD CMP SORT	4. 48-FOLD CMP SORT	4. DBS OPERATOR = 200ms GAP = 4ms
5. DBS OPERATOR = 70ms GAP = 8ms	5. DBS OPERATOR = 210ms GAP = 8ms	5. CMP GATHER
6. VELOCITY ANALYSIS	6. VELOCITY ANALYSIS	6. VELOCITY ANALYSIS
7. 48-FOLD STACK	7. 48-FOLD STACK	7. 48-FOLD STACK tV^2 AMPLITUDE RECOVERY
8. DAS OPERATOR = 200ms GAP = 25ms	8. DAS OPERATOR = 210ms GAP = 16ms	8. DAS OPERATOR = 200ms GAP = 12ms
9. FK FILTER	9. TVF	9. TVF
10. TVF		10. DATUM CORRECTION, +4ms

Figure 5. Seismic processing parameters for relative amplitude stack section though Well 1.

The level of gas measured in the mud returning to the surface is dependent upon drilling rate (ROP) and mud flow rate. In a formation of constant lithology and constant gas concentration, an increase in ROP, combined with a constant mud circulation rate, will produce an increase in the measured gas level because a greater volume of rock per unit time has given up its gas content to the mud fluid. Therefore, the gas levels measured from mud returns at the surface do not relate to the gas concentration in the formation in a simple way.

The reprocessed seismic data showed that the risk of encountering shallow gas at the proposed location was in fact very low. However, four separate potential shallow gas zones had been predicted and the drillers had taken the precaution of drilling a 12¼" pilot with MWD hole before reaming out the borehole to the full 26" diameter.

So, what did we learn from this expensive lesson? Seismic processing of shallow gas survey data is critical and must be monitored very closely. In particular, for the seismic examples shown, we established the following key points from the processing sequences applied (Figure 5);

1. Gain recovery using inverse amplitude decay curves is a data-based, empirical gain recovery procedure which is inappropriate for use pre-stack. It seems to equalise the data and produced amplitudes which were grossly inconsistent at line intersections. In order to recover true amplitude information, a more deterministic or predictive approach is desirable (Hatton et al., 1986).

2. The original deconvolution before stack (DBS) operator was too short (70ms) to attack the water bottom multiple period (162ms) prior to stacking.

3. The care and attention applied pre-stack for the reprocessed section of Figure 2 removed the requirement for post-stack FK filtering applied to the original data of Figure 1.

Simple but effective, theoretically sound pre-stack processing is the key to reliable true amplitude stack sections and is an essential prerequisite to conditioning data for AVO analysis.

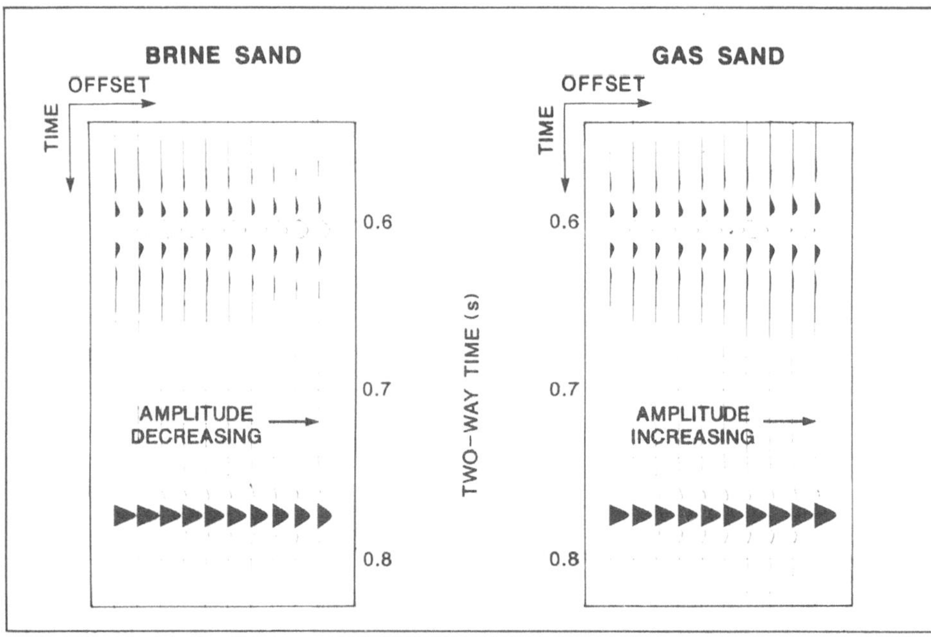

Overlying shale : V_p = 2344ms^{-1}, V_s = 1253ms^{-1}, ρ = 2.43gcm^{-3}, σ = 0.30
Gas sand : V_p = 1900ms^{-1}, V_s = 1266ms^{-1}, ρ = 2.14 gcm^{-3}, σ = 0.10
Brine sand : V_p = 2368ms^{-1}, V_s = 1266ms^{-1}, ρ = 2.14 gcm^{-3}, σ = 0.30

Figure 6. Synthetic NMO corrected, CMP gathers showing AVO effects for brine and gas sand models in two display polarities. (Displays courtesy of Simon Geolithic)

AMPLITUDE VERSUS OFFSET PHENOMENA

Having established an effective pre-stack processing route which delivered apparently reliable true amplitude stacks, we were still conscious of the need to develop more objective criteria to distinguish between those high-amplitude reflectors indicating the presence of shallow gas and those which simply indicated an abrupt change in lithology (cf. coal, peat and lignite layers). We decided to investigate amplitude versus offset (AVO) phenomena.

Seismic amplitudes and AVO effects are discussed by Hilterman (1975), Taner and Koehler (1981), Ostrander (1984) and Shuey (1985).

Figure 6 illustrates the type of effect for which we are looking with AVO analysis of pre-stack, NMO (normal moveout) corrected, CMP (common mid-point) gathers. The significant increase of reflected amplitude with increasing offset range is a direct result of the contrast in Poisson's ratio across the reflecting boundary (Figure 7). Gas-saturated, high-porosity sandstones tend to have abnormally low Poisson's ratios of order 0.10 (Ostrander, 1984). This means that on seismic data which are plotted with the convention that a positive reflection coefficient is represented by a negative number and a corresponding white trough, the top of a gas-bearing sand (assumed for illustration to produce a negative reflection coefficient, which is generally the case) will show a very high amplitude black peak, or "bright spot", on the true amplitude stack. The reflector will show a dramatic increase in amplitude across the CMP gather (Figure 6). The change in amplitude with offset may be so marked that the polarity of the reflector may change from positive at zero offset to negative at the maximum offset range (Figure 7). This latter occurrence may lead to a "dim spot" hydrocarbon indicator on the true amplitude stack.

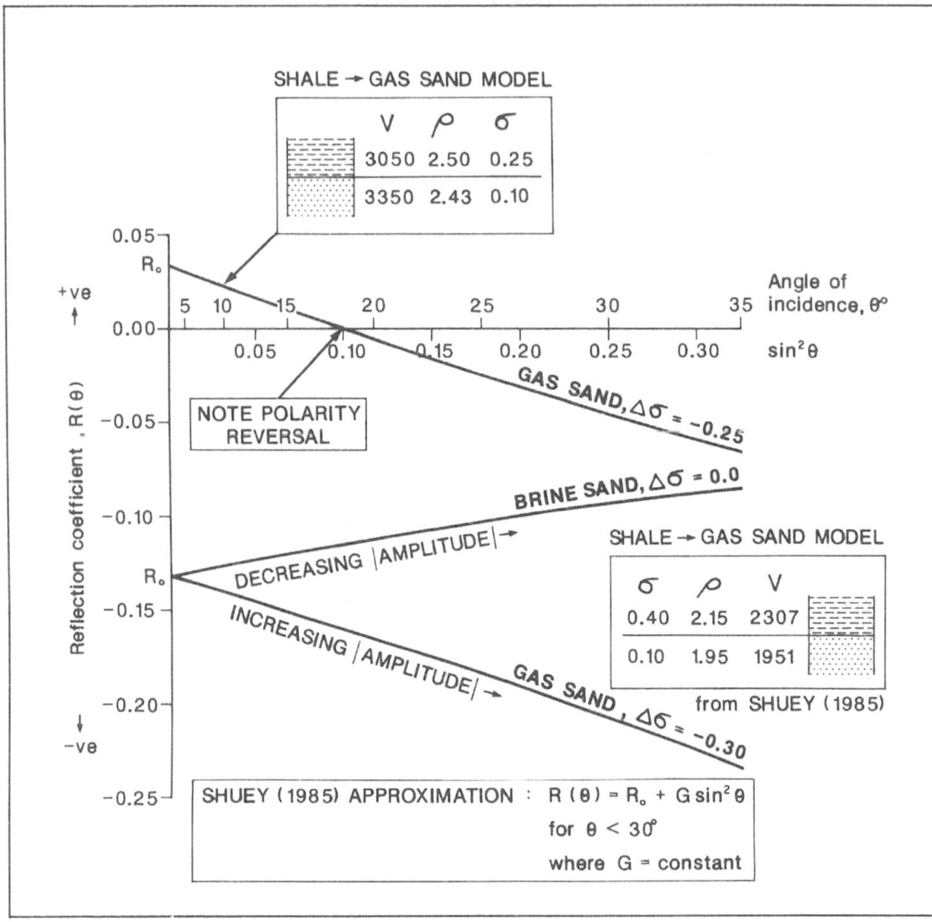

Figure 7. Graph showing AVO effects as a function of Poisson's ratio changes across a reflecting boundary. The curves were generated using the complete simplified equation for $R(\theta)$ of Shuey (1985).

V=interval velocity, ms^{-1} ρ = density, gcm^{-3} σ = Poisson's ratio

WELL TWO

The interpretation of the site survey digital seismic data for this well included an effective analysis of all potential "bright spots" observed on the true amplitude stack data. The objective was to eliminate those high-amplitude reflectors caused by lithological contrast only.

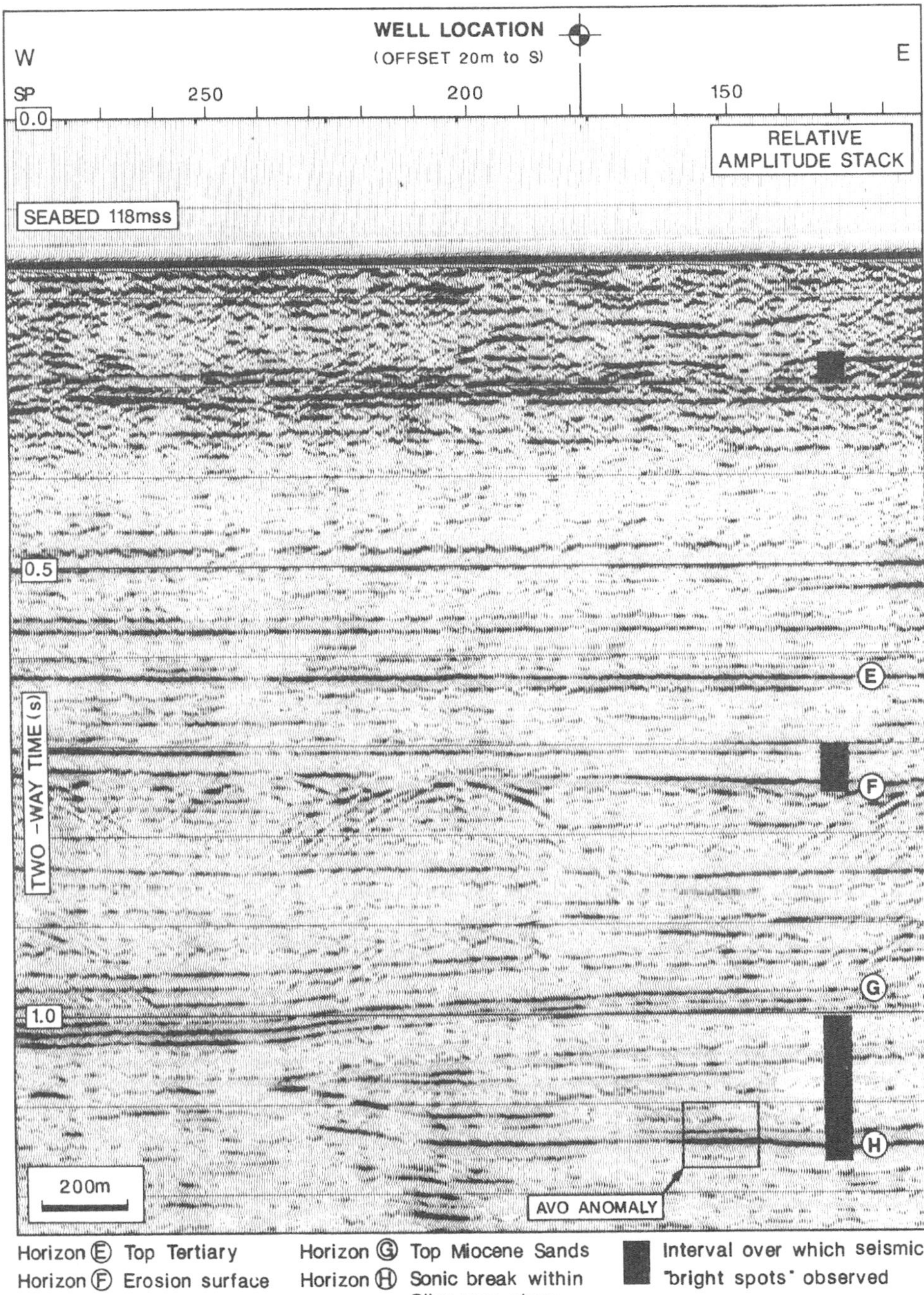

Figure 8. Relative Amplitude Stack through Well Two. The location of the increasing AVO anomaly is shown along the prominent reflector at ca. 1140ms.

Figure 8 shows the relative amplitude stack through well 2. A number of high-amplitude reflectors were identified by the contractor across the survey area and these were grouped into three depth bands; 95-125m, 530-570m and 860-1050m below seabed respectively (these intervals are shown in time on Figure 8).

All but one of the high-amplitude events were dismissed as potential shallow gas accumulations by geological argument, reflector geometry and the absence of any increasing AVO effect.

However, the contractor noted;

> "... one bright spot located 230m east of the location (1050m below seabed) exhibits recognisable amplitude increase with offset. It is therefore considered that this may represent a gas accumulation and appropriate caution is advised."

The increasing AVO anomaly referred to is shown on Figure 9.

The predicted depth, equivalent to 1194mBKB, was in the 17½" drilling phase which would be performed with a riser and blowout preventers.

Figure 10 shows the actual drilling results. The contractor's depth prediction to the Top Miocene sands, Horizon G was very accurate (helped by correlation with a nearby well). Horizon F may correlate with the increase in gamma ray response at 717mBKB; although there is no associated sonic break at this level, there may be a significant density contrast between the siltstone overlying the claystone (no density log run) giving rise to the seismic event observed on the seismic section (Figure 8). The predicted depth of Horizon H matches the base of a mudstone overlying a very silty

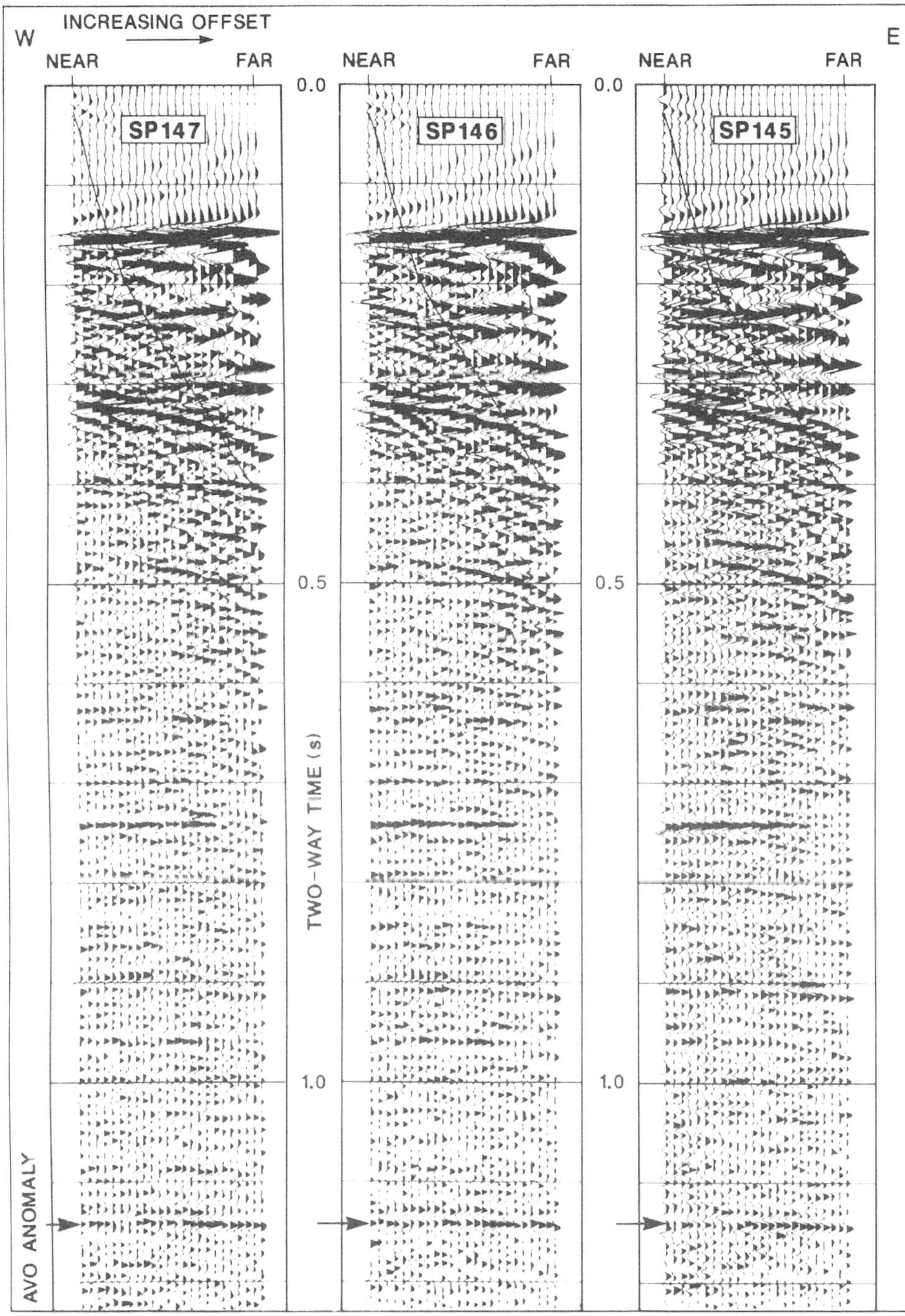

Figure 9. NMO corrected, CMP gathers showing increasing AVO anomaly some 375m east of Well Two at 1140ms.

clay, on Figure 10, by correlation with nearby well data. However, a comparison of the seismic time interval between Horizon G and Horizon H, and the integrated sonic log time interval suggests a revised correlation of Horizon H with the small, negative sonic break at 1166mBKB.

The gas log indicates the presence of gas in the mud returns generally increasing with depth and showing peak values up to 23% between 715mBKB and 856mBKB. This gas is associated with a clay/silty clay interval within which traces of lignite were reported. Despite the relatively high gas levels recorded, this gas is not free to move from the formation due to the clay/silty clay lithology and does not represent a drilling hazard. The background gas level remains fairly constant at 3% from 856mBKB down to 1022mBKB.

Immediately below a 1.5m thick limestone band, the gas level rises abruptly from the background at 3% to a peak of 8%, gradually returning to background levels over the next 73m of sand interval. The top of the Miocene sand at 995mBKB shows no variation in the gas content of the mud returns. The thin limestone interbeds within the sand unit are described on the log as soft to moderately hard and microcrystalline. We suggest that the limestone band has acted as a seal, trapping the gas in the sand interval below it, with the highest concentration of gas immediately below the sealing unit.

This 'gassy' sand interval was not predicted, although it does fall just within the predicted depth range of seismic "bright spots" (Figure 8). We cannot identify a convincing increasing AVO effect at this level (ca.1005ms). The change in interval velocity between the water-bearing sand above the limestone layer (1903ms^{-1}, equivalent to 160μsec.ft^{-1}) and the 'gassy' sand below it (1865ms^{-1}, equivalent to 163μsec.ft^{-1}) is small at about 2%. This is complicated by the presence of the thin limestone (thickness = 1.5m,

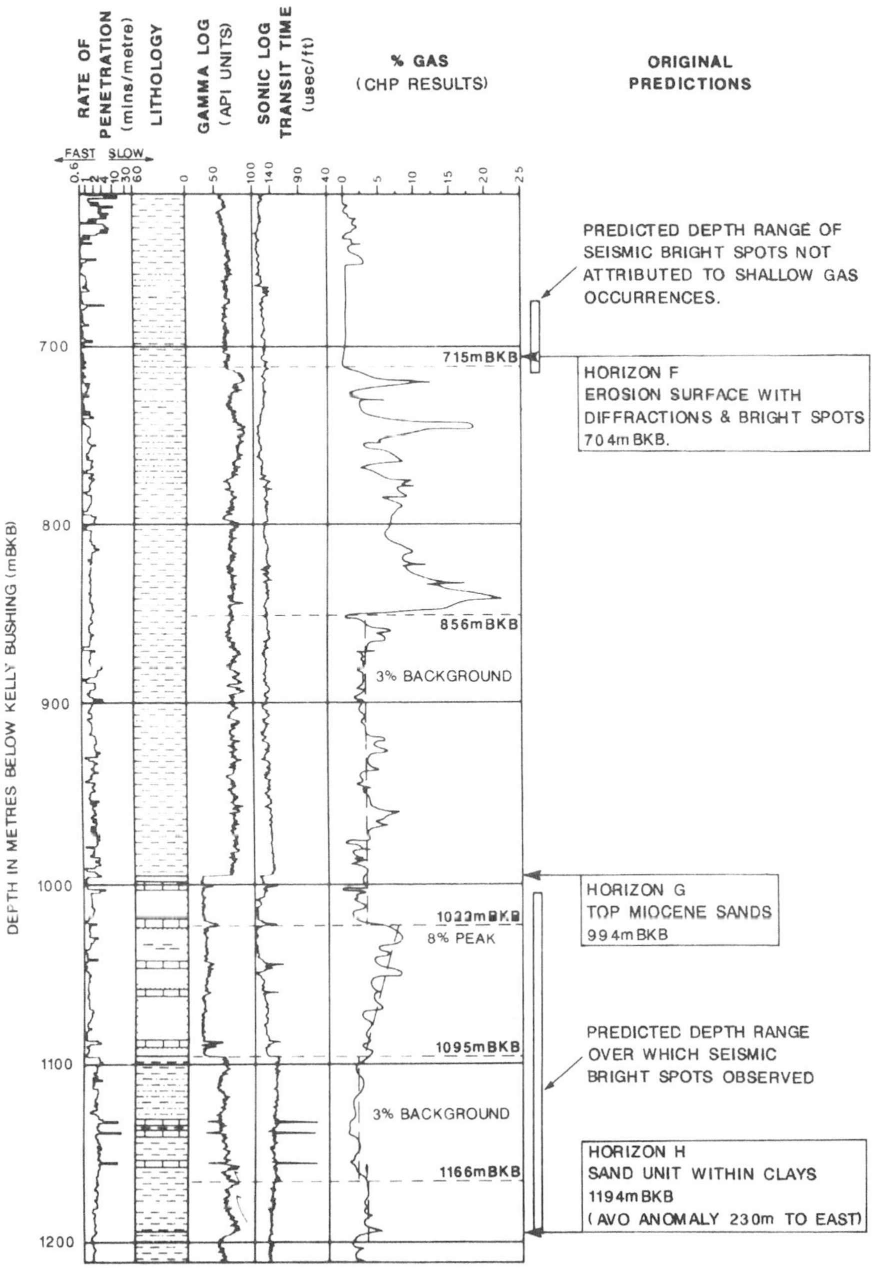

Figure 10. Well 2 : Shallow gas occurrences versus predictions.

interval velocity = $2345ms^{-1}$, equivalent to $130\mu sec.ft^{-1}$) which, due to constructive interference of reflections from its top and base, could yield a seismic event of amplitude 0.80A (A = reflected amplitude from a single finite step) for a dominant frequency at 100Hz (Widess, 1973). There is another seismic event some 25ms below the Top Miocene sand reflector (980ms) on Figure 8 which is consistent with the level of the limestone band. However, we do not fully understand the interaction of these related factors governing the appearance or absence of AVO phenomena at this level but we speculate that they are obscured by the presence of the thin limestone band. Future seismic modelling may shed light on this issue.

The increasing AVO anomaly (Figure 9) identified at Horizon H some 230m to the east of the well location (Figure 8) was not penetrated by the well. However, Horizon H is interpreted to represent a negative sonic break at 1166mBKB which correlates with the top of a coarsening upward sequence on the gamma log (mudstone grading upwards into siltstone) and the onset of a 'gassier' unit (4.0 to 4.5%) on the gas log. This interpretation offers some support for the weak increasing AVO effect observed on the NMO corrected, CMP gathers (Figure 9) by the contractor. The original depth prediction of 1194mBKB is some 28m too deep but still well within the ±5% prediction error of ±58m at this depth.

In conclusion, our analysis shows that the contractor predicted the presence of a 'gassy', "clastic" interval, on the basis of a weak increasing AVO anomaly, to within 2.5% of the depth at which it was encountered in the well. Another 'gassy' sand some 144m shallower was not predicted, but no AVO phenomenon can presently be identified at this level. Seismic modelling may solve this problem.

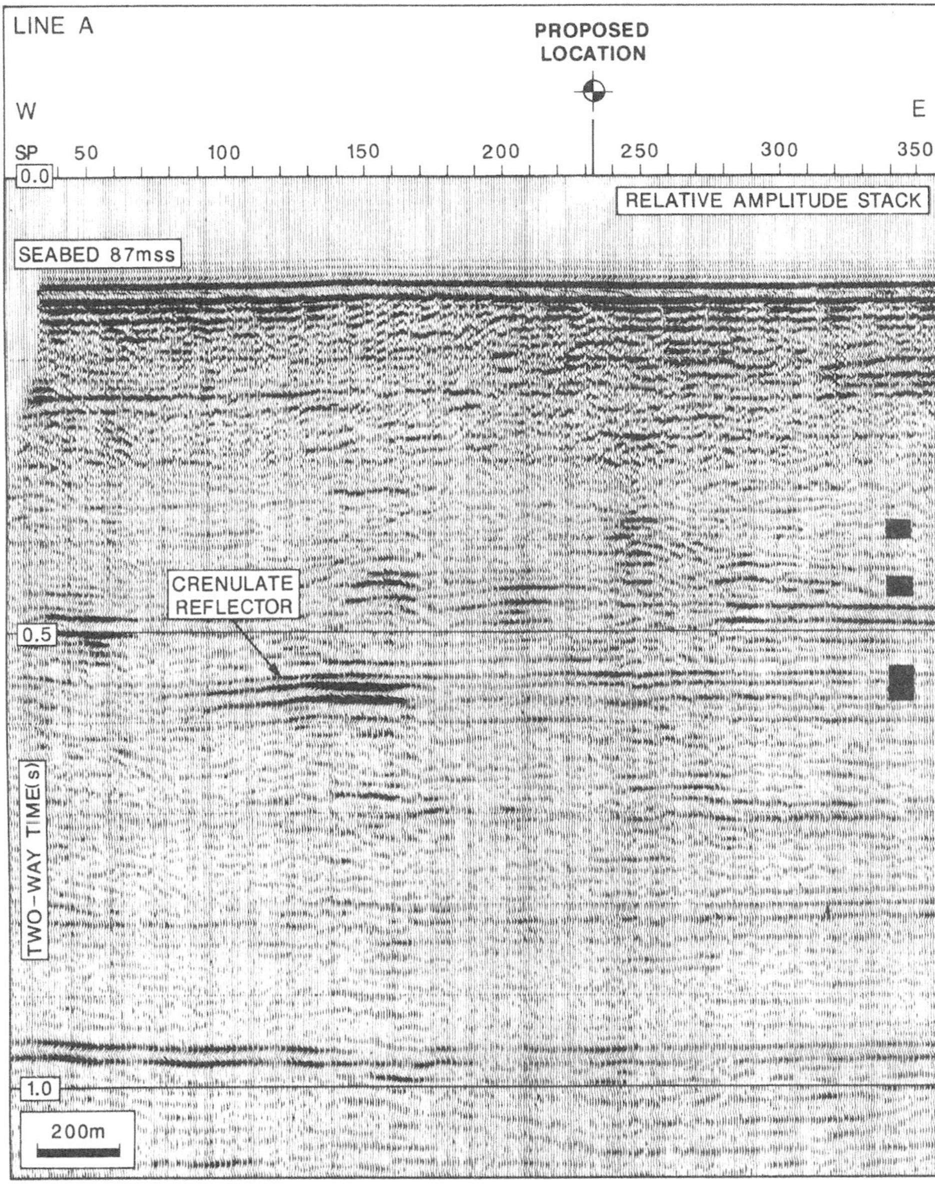

Figure 11. Relative Amplitude Stack through proposed location of Well Three.
Seismic source = watergun (80in^3), 48-fold stack.

The range of predicted potential gas hazards is shown as a vertical bar on the right hand side of the section.

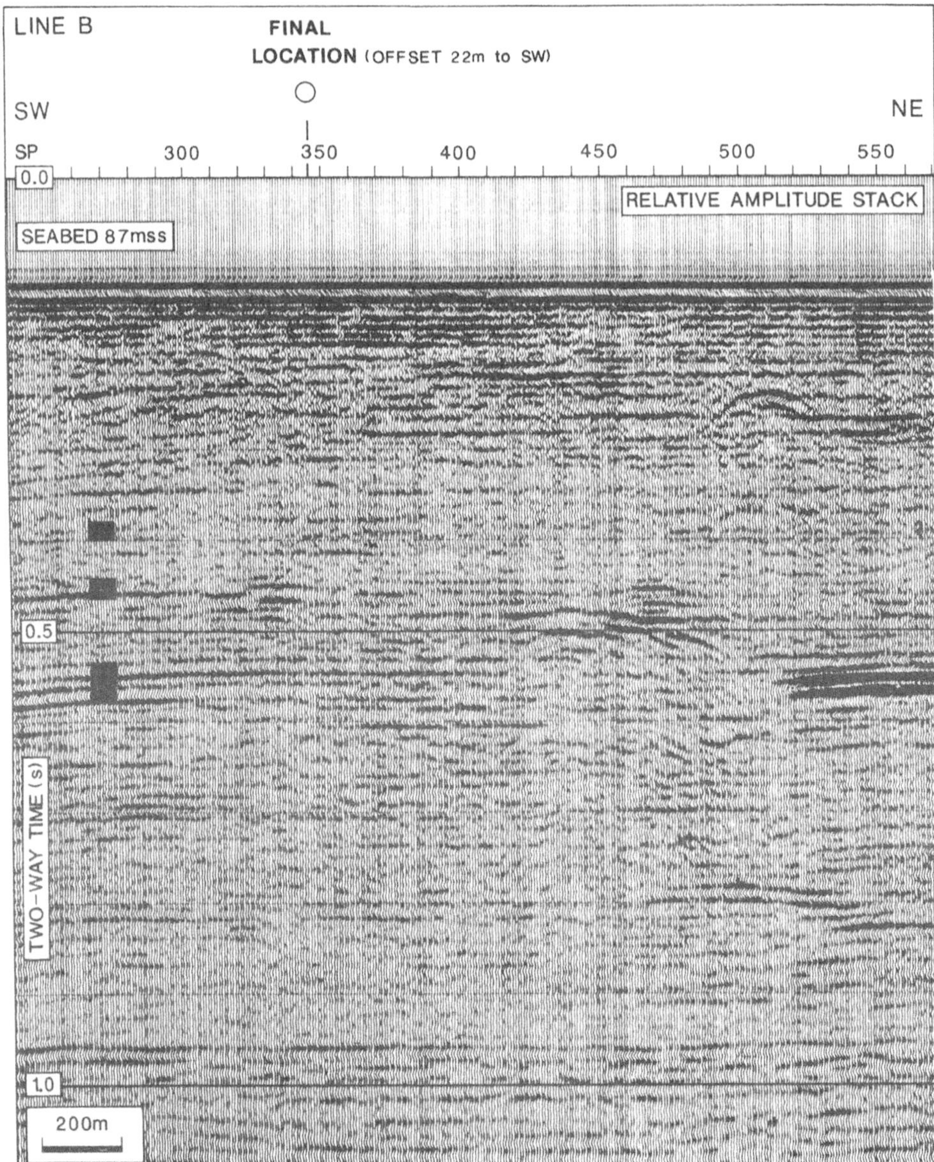

Figure 12. Relative Amplitude Stack through final location of Well Three. Seismic source = watergun (80in^3), 48-fold stack.

The range of predicted potential gas hazards is shown as a vertical bar on the left hand side of the section.

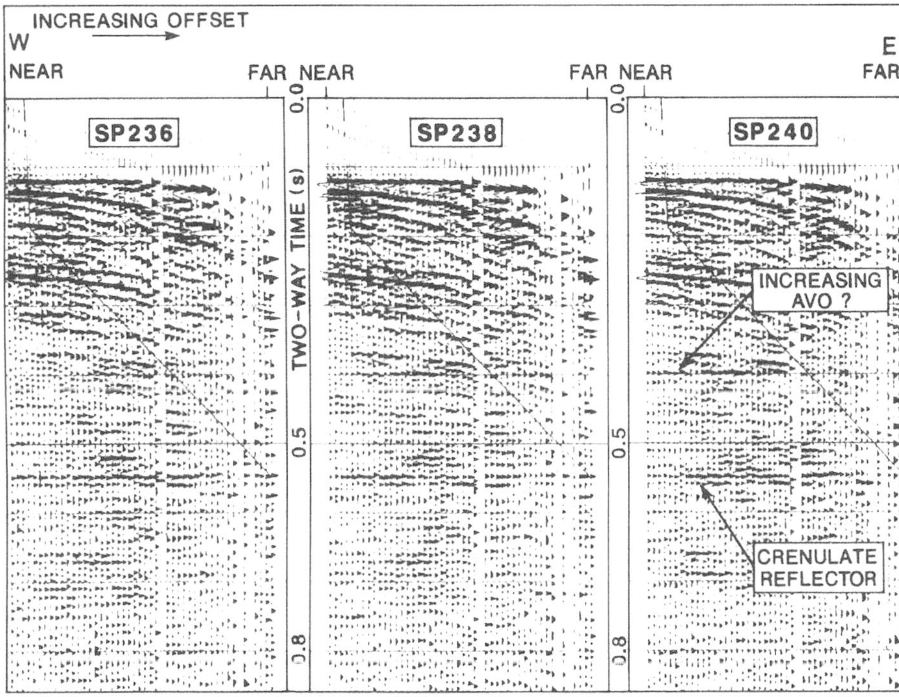

Figure 13. NMO corrected, CMP gathers showing increasing AVO anomaly along Crenulate Reflector (Seismic Line A, Figure 11).

WELL THREE

In the vicinity of well 3, the contractor recognised several laterally discontinuous reflectors of anomalously high amplitude. At the actual location, three potential gas hazards were identified by the contractor: 220 to 245mbsb (380-400ms) some 25m to the E, 290mbsb (450ms) some 35m to the W and 370 to 405mbsb (540-570ms). The latter horizon was widespread and showed very high amplitudes and reversed phase, probably indicative of lignite or gas-charged clastics. The contractor interpreted this horizon as the "Crenulate Reflector" of Holmes (1977), a channelled sand deposited on an unconformity within a clay-dominated sequence.

The proposed location on seismic line A upon which the gas hazard prognosis was based is shown on Figure 11. The final location of the well was some 33m to the N and was situated closer to seismic line B shown on Figure 12. An example of an increasing AVO anomaly along the Crenulate Reflector is shown on Figure 13.

The drilling results are shown on Figure 14. The gamma ray log was acquired through 20" casing and its response is therefore much attenuated, possibly unreliable. No other electric logs are available. An approximate lithology has been inferred from the mud log and the drilling rate (ROP). The gas log shows a consistent increase in the level of background gas with increasing depth which contrasts with a gradual slowing of the rate of penetration over the same depth interval. Since a slower rate of penetration through a constant lithology with constant gas concentration (and a constant mud circulation rate) should give a lower gas reading, we can infer that either the formation is becoming increasingly gas-bearing with depth or there is significant diffusion of gas through the mud cake over a particular gas-bearing zone (or zones) within the open hole section.

The mudlog for a nearby well records very thin sand intervals with associated gas peaks which lends some support to the lithological interpretation proposed on Figure 14.

The contractor's predictions proved to be remarkably good, but incomplete.

Zone 1 (220-245mbsb, equivalent to 333-358mBKB) was predicted to be possibly an isolated, gas-charged sand lens on the evidence of an observed AVO anomaly. Mud losses to the formation and an increase in background gas up to a maximum of 6%, together with a gradual speeding up of the penetration rate, occurred between ca.310 and 330mBKB. The error in depth prediction was about

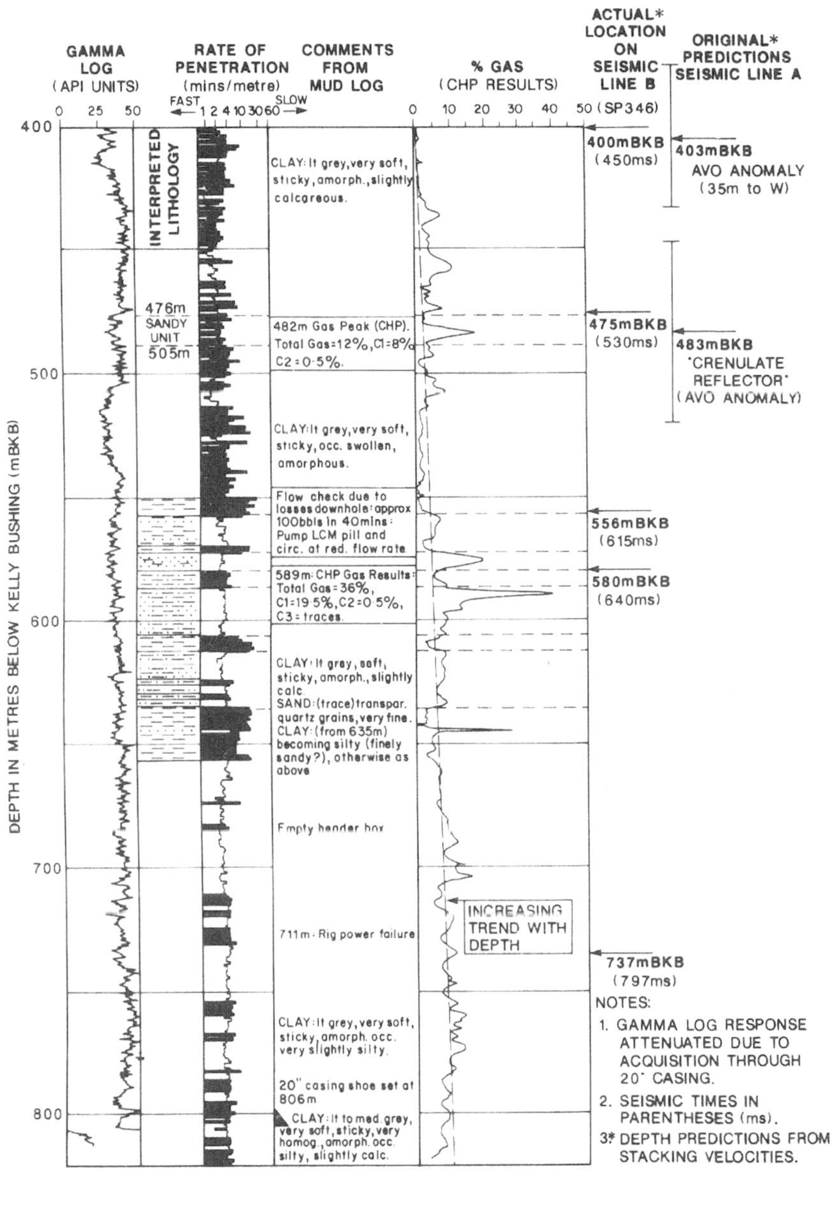

Figure 14. Well 3 : Shallow gas occurrences versus predictions.

7%, but the target is very shallow (which makes % errors rather large).

Zone 2 (290mbsb, equivalent to 403mBKB; Figure 14), despite being a good seismic reflector, appears to have been a false alarm; no drilling break and no increase in background gas.

Zone 3 (370-405mbsb, equivalent to 483-518mBKB) predicted (exactly) the central value of a gas peak of 16%. The "Crenulate Reflector" defining the top of this zone is a very strong, laterally persistent horizon but there is no consistent drilling break. There may be a silty/sandy interval between 476 and 505mBKB based on the gamma log. Alternatively, the interval may simply be a clay/silty clay interval which happens to be relatively rich in carbonaceous material (cf. a similar interval in well 2; Figure 10). The interpretation at this level is ambiguous.

However, the original gas hazard prognosis completely missed the 'gassy' interval between 556mBKB and 635mBKB (Figure 14). The equivalent seismic interval is 615ms to 702ms. The relative amplitude stacks of line A (Figure 11) and line B (Figure 12) show little seismic energy throughout this zone. However, the equalised stack of line B (Figure 15) shows weak reflectors of opposing dip defining mound-like features within the interval. (These dips have been further enhanced by displaying our seismic data at a horizontal scale of 1:10,000 yielding a vertical exaggeration of about 2:1. This simple technique allows low sedimentary dip to be identified and diffraction events to be distinguished from primary events.) The predicted depth at the well location of 556mBKB to the top of such a mound (615ms) coincides with an increase in the gas level to 7% and a dramatic drilling break (from clay/shale to silt/sand). Subsequent gas peaks of 19%, 38% and 8% correlate directly with intervals interpreted to be more silty (This is our preferred interpretation but an ambiguity exists in that a faster drilling rate will

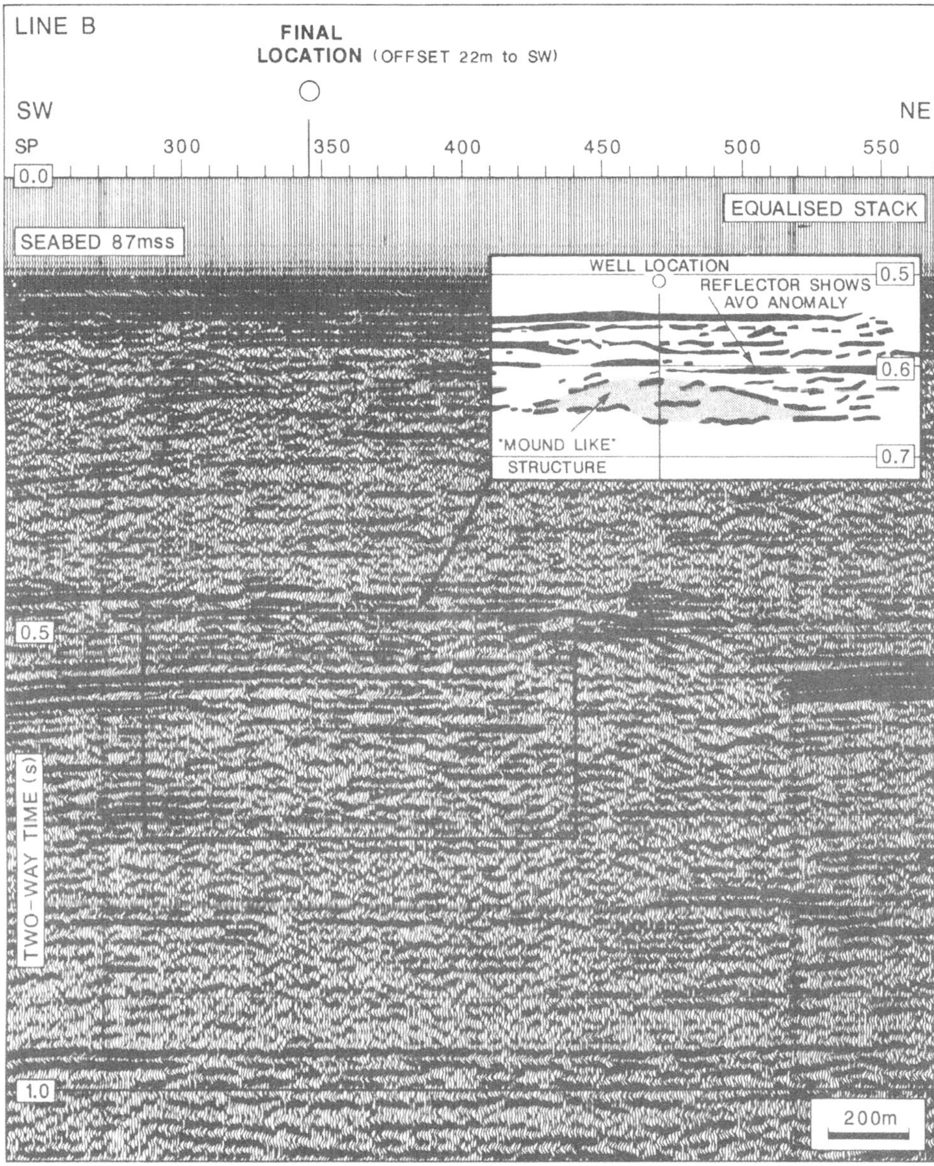

Figure 15. Equalised Stack Through Final Location of Well 3 (Seismic Line B). Inset shows line interpretation of "mound like" feature defined by reflections of opposing dip and penetrated by well. Note flat strong event at 605msTWT which shows increasing AVO effect.

produce higher gas levels in the mud returning to the surface. See discussion of Well 1).

A strong seismic reflector at ca.605ms, about 110m NW of the well location on line B (Figure 15), shows an increasing AVO effect (Figure 16). This reflector continues weakly to the SE and through the well location at approximately the expected seismic time to correlate with the top of the 'gassy' interval. The seismic data suggest an interval of mound-like features which may represent overlapping channel systems producing a vertical sequence of interbedded sands and shales. The shales provide local seal to isolated sand bodies which show different gas concentrations because they are not in communication with one another.

One aspect of conditioning seismic data for AVO analysis concerns us at present. Multiple periodicity remains constant only for vertical incidence and zero-offset recording (Yilmaz, 1987). Therefore, predictive deconvolution, which relies on constant periodicity to remove multiples, may not perform effectively on CMP gathers where multiple periods change with time and offset. Also, spherical divergence compensation using primary stacking velocity information distorts the amplitudes of multiples on non-zero offset data (Yilmaz, 1985). Predictive deconvolution is therefore most effective on the near offset traces. Is it possible that attenuation of seabed multiples on near offset traces, due to a predictive deconvolution which gradually loses effectiveness as the offset increases, could give rise to apparent increasing AVO anomalies? (See Figure 16).

We are currently investigating the slant stack domain, in which the periodicity and amplitudes of multiples are preserved (Yilmaz, 1985), for pre-stack predictive deconvolution and removal of seabed multiples.

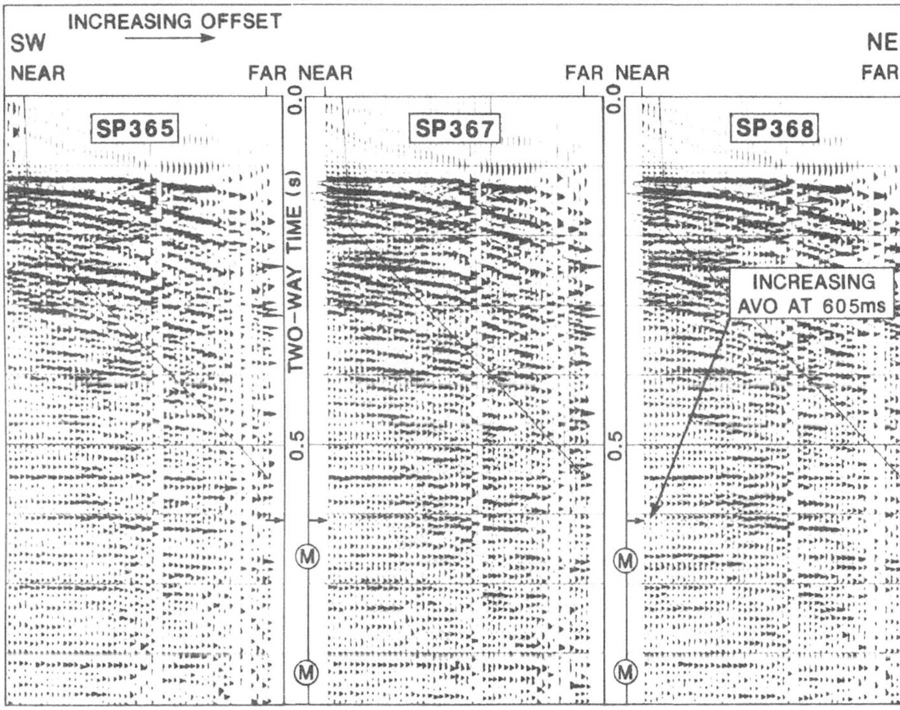

Figure 16. NMO corrected, CMP gathers showing increasing AVO anomaly at time 605ms some 130m to the NE of Well Three. (M) = ?Multiple showing apparent increasing AVO effect.

Below this 'gassy' interval, the background gas steadily increases from about 6% at 635mBKB to perhaps 10% at 800mBKB without any major drilling breaks or significant gas peaks.

In conclusion, the contractor's gas hazard predictions for well 3 were a mixture of success and failure. Missing the main 'gassy' clastic interval was disappointing but very instructive; AVO analyses are only useful if we can identify them in the first place and using all available data is essential. In this case, the equalised stack gave a real clue about reflector geometries which could have led to the recognition of the critical AVO anomaly (easy with hindsight).

CONCLUDING DISCUSSION

Shallow, gas-bearing sands pose a potential danger to drilling operations as hydrostatic pressure cannot be guaranteed to contain the gas due to the inherently weak formation, nor can the gas be constrained by conventional well control equipment set for surface casing for the same reason.

High-resolution, shallow seismic data for the detection of shallow gas hazards must concentrate on the very shallowest part of the section. But this is where fold of cover is lowest due to near surface refraction and persistent water-bottom multiples.

Our studies, in conjunction with the above observations, have highlighted important areas for the reliable prediction of shallow gas accumulation:

1. The need for good quality acquisition using a reliable source with a minimum phase output, as short a near trace offset as possible and the need to increase the number of traces within the shortest offset range (for effective AVO analysis in the shallowest part of the seismic processing);

2. Simple, theoretically sound seismic processing, especially the application of more effective multiple attenuation pre-stack:

3. The desirability of an automated procedure to search for AVO anomalies and to display them in an easily interpreted form;

4. The importance of effective feedback (especially to the contractor responsible for interpretation) of the actual borehole results compared with pre-drilling predictions of shallow gas occurrence;

5. The critical importance of the final scales of display of the seismic sections to enable reliable interpretation.

6. The need for a more consistent reporting scheme in which potential gas hazards are quoted in depth, metres subsea (mss) and seismic time, ms.

7. The critical importance of using all available data and not relying on one technique or display.

TOTAL OIL MARINE p.l.c. currently apply AVO processing to all seismic events which show high amplitudes after true relative amplitude recovery. As our experience of predicted versus actual drilling results grows, we believe that AVO processing may lead eventually to safer drilling practice and reduced drilling costs without any compromise on safety.

ACKNOWLEDGEMENTS

We would like to thank the following companies for permission to publish the seismic and well data in our paper; BP Exploration, Elf UK plc, Marathon, Norsk Hydro and Occidental. Thanks also to Mike Hall and Adrian Pelham of Simon-Geolithic, Swanley and Richard Orren and Phil Meaden of Wimpol Ltd, Swindon for their assistance with data processing and displays. Thanks to Neville Taylor and Tony Rochester who kindly reviewed the text and made significant improvements. And a special thankyou to Alison Sievewright for her patience in typing the text and to Lynne Urquhart, Mary Anderson, Karl Howard and Euan Robb for draughting the figures. Finally, we thank Total Oil Marine p.l.c. for permission to publish our paper and stress that the views expressed are entirely those of the authors and they do not necessarily represent those of the company.

REFERENCES

Desbrandes, R. (1985). Encyclopedia of Well Logging, Graham and Trotman, London.

Hatton, L., Worthington, M. H. and Makin, J. (1986). Seismic Data Processing: Theory and Practice, Blackwell Scientific Publications, Oxford.

Hilterman, F. J. (1975). Amplitudes of Seismic Waves - A Quick Look, Geophysics, 40, 745-762.

Holmes, R. (1977). Quaternary Deposits of the Central North Sea, 5. The Quaternary geology of the UK sector of the North Sea between 56° and 58°N, Rep. Inst. Geol. Sci., 77/14.

Ostrander, W.J. (1984). Plane-wave reflection coefficients for gas sands at nonnormal angles of incidence, Geophysics, 49, 1637-1648.

Shuey, R. T. (1985). A simplification of the Zoeppritz equations, Geophysics, 50, 609-614.

Taner, M. T. and Koehler, F. (1981). Surface consistent corrections, Geophysics, 46, 17-22.

Widess, M. B. (1973). How thin is a thin bed?, Geophysics, 38, 1176-1180.

Yilmaz, O. (1987). Seismic Data Processing, Society of Exploration Geophysicists, Tulsa, Oklahoma.

SHALLOW GAS RESERVOIRS AND MIGRATION PATHS OVER A CENTRAL NORTH SEA DIAPIR

R. S. K. Salisbury

Britsurvey

ABSTRACT

Studies of relative amplitude processed sections over a Central North Sea Diapir structure have identified shallow gas reservoirs within Quaternary and Tertiary sediments in the interval up to 600 metres below seabed. They are believed to have formed within silt and sand layers in an otherwise clay sequence.

Two pathways for the migration of thermogenic gas into the shallow section are identified, namely Tertiary fault planes and vertical columns of gassified sediment, or (seismic) gas chimneys.

Migration of gas into individual reservoirs has taken place either vertically to a seal or partial seal, or horizontally along silt and sand layers, away from the vertical migration

Volume 25: Safety in Offshore Drilling, 167–180.

paths. This has resulted in differing seismic responses from the upper and lower limits of the reservoirs.

Estimates of the partial pressure exerted by the gas column at the structural high point, and the volume of gas in place have been made for selected individual reservoirs.

INTRODUCTION

High resolution seismic data was acquired by Britsurvey over a diapir structure in the Central North Sea, with the prime objective of identifying any potential shallow gas reservoirs and assessing them as drilling hazards. Details of the equipment spread are shown in Figure 1 together with the acquisition parameters. Line spacing was 100 metres increasing to 50 metres in some places, and all lines were migrated during processing.

ACQUISITION	
SOURCE:	4 x 40 cu. in. Sleeve Airgun Cluster
STREAMER:	Programmed to 96 chs. 6.25 m groups Linearly spaced 'phones
RECORDING SYSTEM:	TTS 128, Sample rate 1ms, Record length 1.5 secs
SHOT INTERVAL:	6.25 m
LINE SPACING:	50 or 100 m

PROCESSING
(Salient Points)

- 48 fold CDP sort, 6.25m s.p. interval, 3.125 CDP interval
- DBS using 40 msec operator, 8 msec gap
- DMO test run but not applied because of time constraints
- DAS using 240 msec operator, 18 msec gap
- f/k migration
- Relative amplitude and Equalised display

Figure 1. Summary of Acquisition and Processing Parameters

The data, which was interpreted in conjunction with exploration seismic lines and five well logs from within the area of coverage, shows a late Cenozoic sequence of clays and claystones with subordinate silt and sand bodies in the form of lenses and layers, typical of the Central Graben of the North Sea. In this case, however, movement of the underlying diapir had caused doming and halokinetic fracturing of the formation. Figure 2 is a section which shows the main structural features identified.

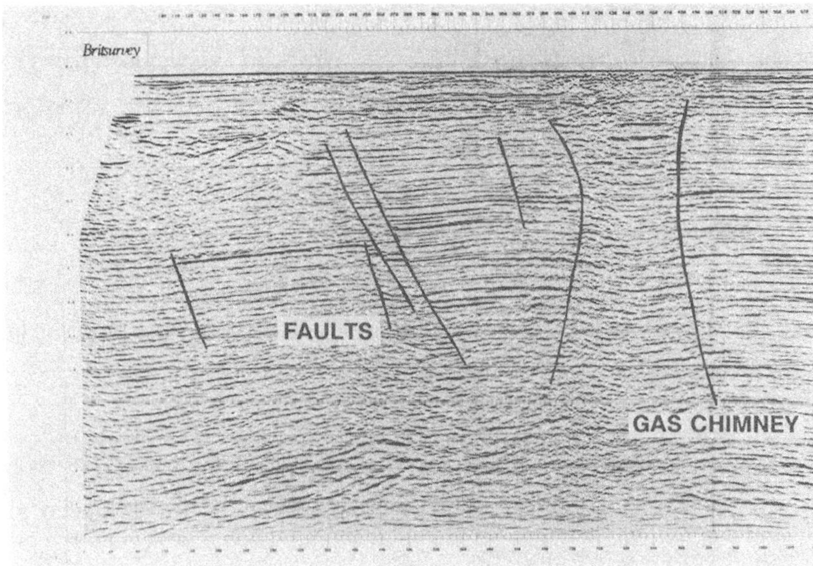

Figure 2. *Equalised Section Through Study Area Showing Main Structural Features Including Faulting and Gas Chimney*

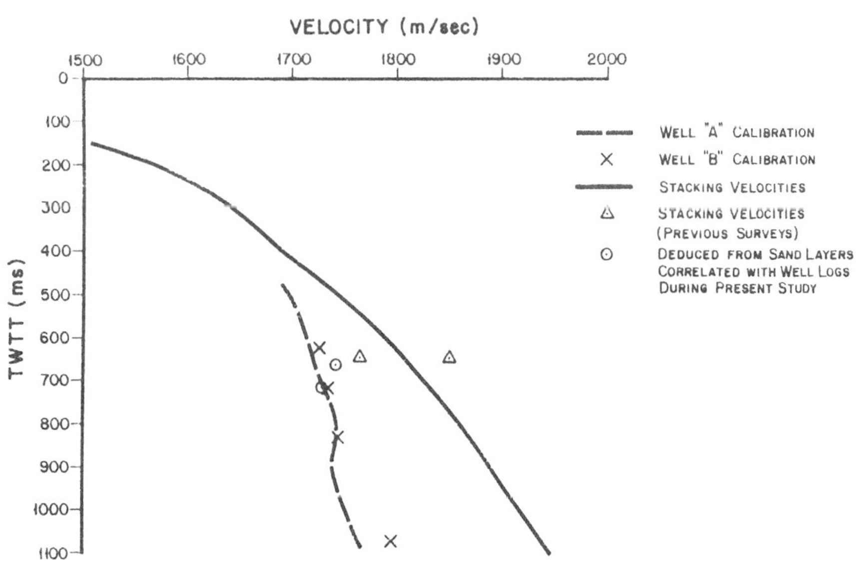

Figure 3. *Comparison of Velocities, Showing Difference Between Stacking Velocities and Well Calibration Velocities.*

The study identifies shallow gas reservoirs in several contrasting geological settings, and highlights variations in their seismic attributes in relation to the migration paths involved in their formation. Estimates of the pressure and volume of gas within individual reservoirs have been made for comparison purposes.

DEPTH CONVERSION

In this case particular importance was given to depth conversion. Velocity calibrations from two of the wells were compared with an average velocity function derived from the stacking velocities used in processing (Figure 3).

A velocity profile was subsequently derived based mainly on the well velocity calibration, although since 375 metres below seabed is the upper limit of this calibration, stacking velocities were used above this depth.

INTERPRETATION

Data interpretation has concentrated on the upper 800 milliseconds TWTT of data (corresponding to just over 600 metres below seabed), where there are a large number of amplitude anomalies, many of which are interpreted as shallow gas reservoirs (Figure 4).

The most extensive amplitude anomalies are associated with Reflector 5 (Figure 5), which has been confirmed as a major sand layer through well correlation. Amplitude anomalies extend from high points on the reflector at about 450 metres below seabed, to points between 490 and 500 metres below seabed in a radial pattern corresponding to the structural form of the area.

Most other amplitude anomalies are localised features on reflectors which, through careful correlation with well logs, are identified as silt and sand layers, and some fall on uncalibrated reflectors presumed to represent either less well formed silt and sand layers or laterally discontinuous lenses, neither of which register in the well logs.

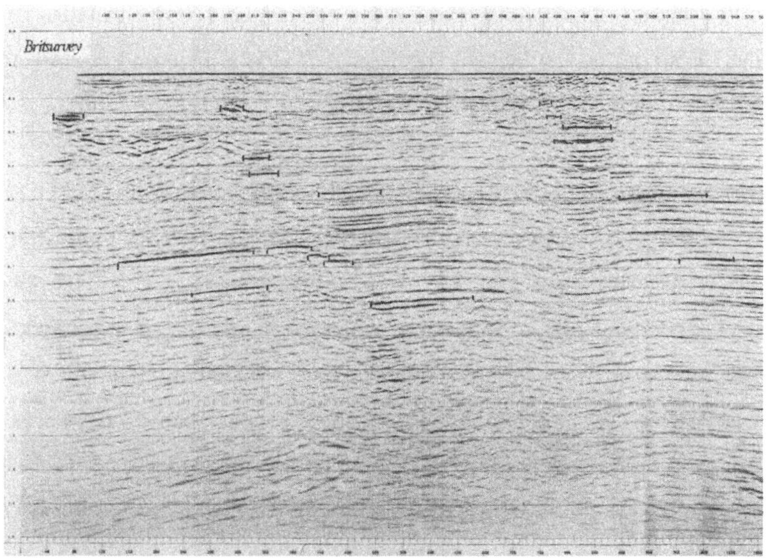

Figure 4. Relative Amplitude Section Through Study Area Showing Amplitude Anomalies Associated with Faults and Gas Chimney

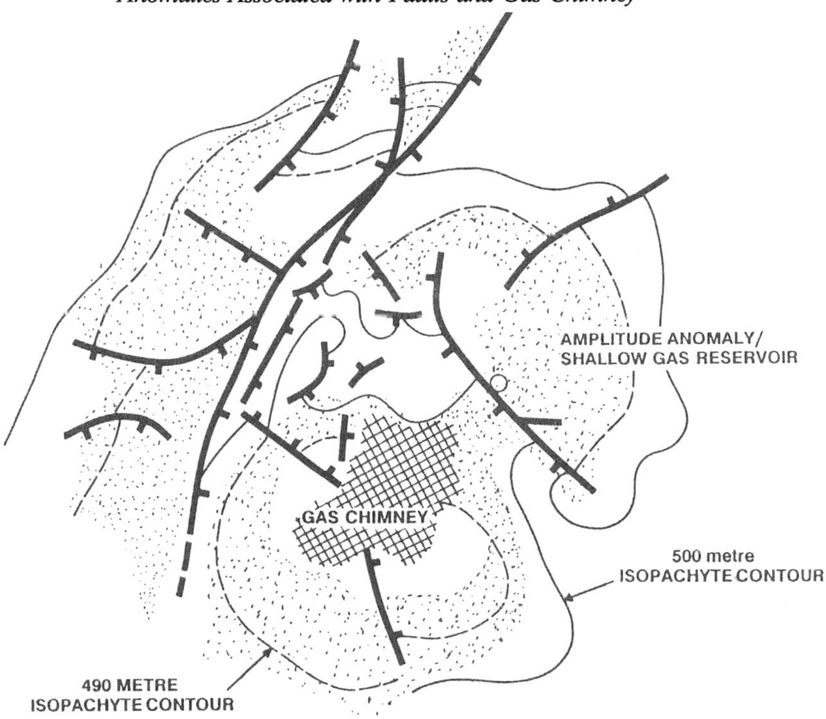

Figure 5. Map of Reflector 5 Showing Isopachyte Contours, Fault pattern, and Amplitude Anomalies/Shallow Gas Reservoirs

Most of the higher amplitude anomalies are associated either with faults or with gas chimneys. Both of these are clearly resolved on the data (Figure 4) and are considered to be feasible routes for the migration of gas from a thermogenic source.

The faults are clearly resolved to a depth of about 800 milliseconds TWTT and extend upward to within 100 metres or so of the seabed. They are normal faults with throws of up to a few tens of metres, and are almost certainly the result of a tensional regime caused by updoming of the underlying salt diapir. Adjacent to the faults amplitude anomalies occur at many different levels (Figure 4), although they are particularly numerous between 400 and 600 milliseconds TWTT. Most extend only a short distance away from the fault into the formation, although others, in particular Reflector 5, are extensive. Amplitude anomalies are also present within Quaternary channel fill sediments, above the upper limit of vertical displacement by the faults.

There are also strong indications from analogue seismic and side scan sonar data above the major faults, that gas charged sediment is present immediately below the seabed and that seabed gas seeps are present (Figure 6). This suggests that the faults are active migration paths for shallow gas. Particularly well defined reflective patches up to 100 metres long on the side scan sonar data may represent the remains of a rich growth of marine life stimulated by the presence of hydrocarbon bearing sediment (Hovland and Sommerville, 1985).

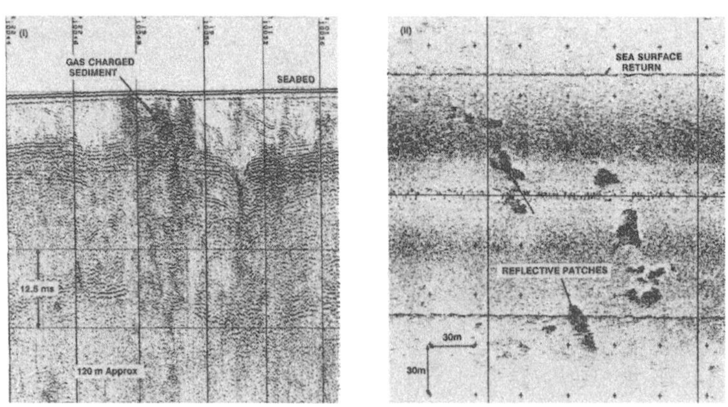

Figure 6. *Analogue Seismic (i) and Side Scan Sonar (ii) Records Showing Gas Charged Sediment at Seabed Above Fault*

Gas chimneys are identified by the characteristic zone of poor seismic data extending over a considerable vertical distance and exhibiting a high degree of energy scattering, poor definition of shallow reflectors and the masking of deeper data. Strong amplitude anomalies are present at all levels and particularly in the shallow part of the section, although the extensive masking of deeper anomalies is likely. Vertical movement of gas is thought to have taken place by disseminating migration, possibly via microfissuring of clay. Again, there is evidence from analogue seismic and side scan sonar data (Figure 7) of gas charged sediment immediately below the seabed, which strongly suggests that gas chimneys are also active vertical migration paths for shallow gas.

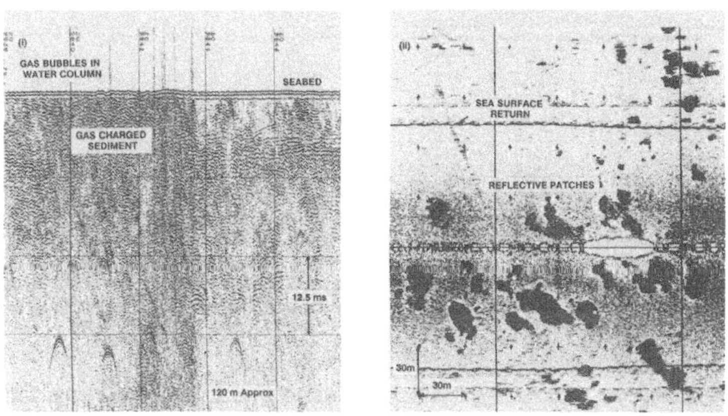

Figure 7. *Analogue Seismic (i) and Side Scan Sonar (ii) Records Showing Gas Charged Sediment at Seabed Above Gas Chimney*

SEISMIC ATTRIBUTES OF SHALLOW GAS RESERVOIRS

With feasible migration paths established, a large number of the amplitude anomalies are considered to be shallow gas reservoirs. Close examination of individual amplitude anomalies reveals differences in seismic response and two categories of shallow gas reservoirs, Category A and Category B have been identified (Figure 8). At the upper limit Category B reservoirs typically show a clear, phase reversed reflector, unlike Category A reservoirs which typically show a less strong reflector in which phase inversion is more difficult to establish. At the lower limit Category A reservoirs typically exhibit a strong, (normal phase) basal reflector, but with Category B reservoirs this is typically poorly defined.

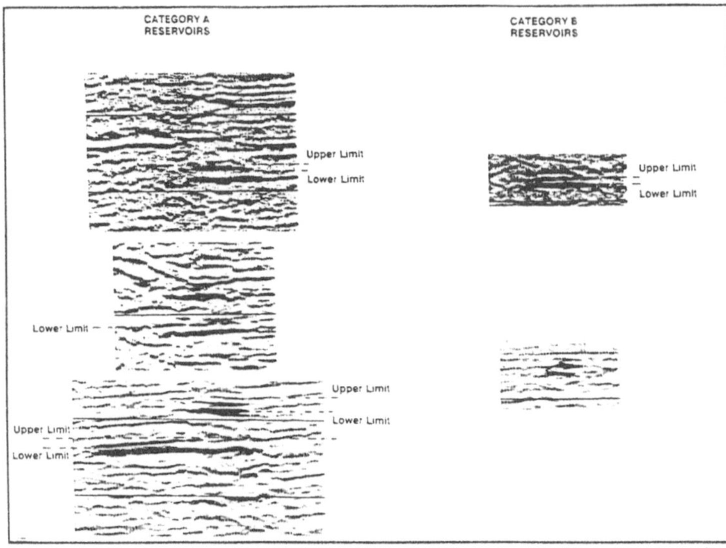

Figure 8. Examples of Category A and Category B Shallow Gas Reservoirs Illustrating Differences in the Seismic Response of the Upper and Lower Limits of the Reservoirs. In Category A Reservoirs Migration of Gas Has Taken Place Horizontally and in Category B Reservoirs Migration of Gas Has Taken Place Vertically

Category A reservoirs are mainly, but not exclusively, associated with the faults, from which they extend laterally into the surrounding sediments for distances of up to a few hundred metres, remaining conformable to the bedding. Most of the Category B reservoirs are within the gas chimneys, and they predominate within the upper part of the section, although the extensive masking of deeper data is thought likely. The shallow reservoirs within Quaternary channel fill sediments above the faults may also be considered Category B reservoirs.

Category A reservoirs are those which have been filled by the horizontal migration of gas along silt and sand layers. We may expect sediments immediately below Category A reservoirs to be largely free from gas, thus the basal interface constitutes a transition from gassified sand to clay, producing the strong normal phase reflector observed.

In contrast, Category B reservoirs are those which have been filled by the vertical migration of gas to a seal or partial seal. The poorly defined basal reflector of the Category B reservoirs is attributed to quantities of gas within the migration path immediately below, which preclude a clear interface between gassified and non-gassified sediment.

The contrast between the upper limit reflectors of the Category A and the Category B reservoirs is less well defined, but is explained as follows. The clear, phase reversed reflector of the Category B reservoirs is attributed to the seal of partial seal to vertically migrating gas, which would produce a clear, clay to gassified sand interface. It would appear that this is not the case with Category A reservoirs, where the lack of a clear interface is possibly due to infusion of gas upward from the reservoir.

CALCULATION OF PRESSURES AND VOLUMES

A quantitative estimation of the overpressure of gas due to the density contrast between water and gas, and the volume of gas in place was made for selected shallow gas reservoirs, as outlined in Figure 9. Calculation of gas pressure was based on the procedure described by Isaksen et al (1987). Table 1 shows some examples of the results obtained, although there is some doubt as to the validity of the absolute values. This is primarily because of uncertainties of establishing the upper and lower limits of reservoirs (and thus the height of the gas column in place) from the seismic records, and also in determining the thickness of individual reservoirs from the well log correlation. Uncertainties associated with the estimation of the respective densities of pore water and gas, and of reservoir porosity are thought to be less significant. Despite these shortcomings, the results are considered to be useful for comparison purposes.

ESTIMATION OF GAS PRESSURES AND VOLUMES

1. Vertical column of gas estimated from seismic data by measuring minimum depth to top (H1), and maximum depth to base (H2) of each anomaly.
2. Assumptions made of the respective densities of pore water (pW) and gas (pG). These are as follows:
$$pW = 1.025 \text{ g/cu. cm.}$$
$$pG = 0.2 \text{ g/cu. cm.}$$
3. The overpressure P in bars for each reservoir calculated as follows:
$$P = \frac{(H2-H1)(pW-pG)}{10}$$
4. Areal extent of each reservoir A measured from seismic data.
5. Thickness of silt/sand layer x estimated using seismic data in conjunction with well logs. (In cases where this was greater than (H2-H1) then P was recalculated using the higher value.)
6. Porosity of unconsolidated sandstone assumed as 40%.
7. Volume of gas at a pressure of 1 bar calculated for each reservoir as follows:
$$V = Ax(P+1)\frac{}{100} \text{ cubic metres}$$

Figure 9. *Summary of Steps Involved in Estimating Pressures and Volumes of Gas in Individual Reservoirs*

ANOMALY REF.NO.	MIN DEPTH BELOW SEABED (metres)	DEPTH RANGE (metres)	OVERPRESSURE P(bars)	AREA (SQ. m.)	SAND LAYER THICKNESS (metres)	VOL OF GAS (S.CU.m.)	REMARKS
57	275	8	0.66	420000	10	2788800	(1)
58	275	13	1.0725	25000	5	103625	
59	259	6	0.495	45000	5	134550	
60	258	6	0.495	75000	5	224250	
61	303	2	0.165	10000	5	23300	(1)
62	245	9	0.7425	10000	8	55760	
69	284	2	0.165	22000	2	20504	
72	290	20	1.65	280000	5	1484000	(2)
73	290	3	0.2475	100000	5	249500	(1),(2)
74	286	7	0.5775	35000	5	110425	
78	225	17	1.4025	120000	10	1153200	(2)
81	313	17	1.4025	450000	5	2162250	(2), DEEPER PORTION OBSCURED
82	390	3	0.2475	15000	2	14970	
83	368	12	0.99	30000	3	71640	
85	313	7	0.5775	35000	6	132510	
93	265	11	0.9075	10000	5	38150	
94	324	20	1.65	40000	5	212000	(2)
95	318	7	0.5775	35000	5	110425	
99	505	4	0.33				AREA UNCERTAIN - VERY SMALL
100	465	47	3.8775	140000	12	3277680	COMPLEX - CROSSED BY FAULTS
101	450	50	4.125	1070000	12	26322000	EXTENSIVE - STRADDLES MAIN SW/NE FAULT
102	505	7	0.5775	35000	12	265020	(1)
103	455	41	3.3825	710000	12	14935560	
104	492	13	1.0725	40000	12	397920	
105	492	6	0.495				AREA UNCERTAIN - VERY SMALL
107	472	10	0.825	10000	12	87600	
111	470	7	0.5775	10000	12	75720	

NOTES (1) SAND LAYER INDICATED TO BE THICKER IN WELL LOG THAN IS SHOWN BY SEISMIC DATA
(2) AMPLITUDE ANOMALY MAY BE MORE EXTENSIVE THAN IS REVEALED BY PRESENT SURVEY
ASSUMED DENSITY OF PORE WATER: 1.025 g/cc
ASSUMED DENSITY OF GAS : 0.2 g/cc
ASSUMED POROSITY OF UNCONSOLIDATED SANDSTONE: 40%

Table 1. *Extracts From Table Showing Estimated Pressures and Volumes of Gas in Individual Reservoirs*

CONCLUSIONS AND LIMITATIONS

Well over a hundred amplitude anomalies were mapped, plotted and their attributes studied in order to assess them as shallow gas reservoirs. The complex fault pattern was mapped at one level, and the relationship of the faults to shallow gas reservoirs throughout the section was examined.

Conventional paper displays of equalised and relative amplitude processed sections were used exclusively during this study, which was extremely labour intensive. For example, mapping of the amplitude anomalies was carried out in several stages, each of which involved examining the anomaly on most lines on which it appears. The limits of the anomaly would then be tied-in with increased certainty at each stage. The fault pattern was both complex and fragmented (Figure 5) with many short faults having small throws, which were nevertheless important in controlling shallow gas reservoirs and migration paths.

Migration of the seismic sections had a dramatic effect in improving the clarity of the data, and accurate mapping of the faults would not otherwise have been possible. A large vertical display scale and the close line spacing were also essential, although difficulties in aligning certain faults where the line spacing was 100 metres may have been resolved by extra lines. The potential for accurate mapping using high resolution data was demonstrated, but proved laborious and time consuming. An automated interpretation workstation may well have enabled comparisons to be made more easily between seismic lines, and would undoubtedly have made the mapping of the faults and amplitude anomalies more efficient.

Despite the high quality of the seismic source and careful quality control of processing, the upper and lower limits of shallow gas reservoirs were not always easily determined and there was some uncertainty in establishing phase inversion. The ability, offered by a work station, to look at expanded scale displays of amplitude anomalies and also to look more closely at the phase and frequency content of the data, possibly in conjunction with AVO information would have been beneficial.
Finally, the inaccuracy of the stacking velocities below 375 metres, and the benefit of having well calibration velocities available, has been illustrated. This is clearly of considerable importance, where the accurate prediction of depths to shallow gas reservoirs must be made.

REFERENCES

1. Badley, M.E. (1985), Practical Seismic Interpretation, D. Reidel Publishing Company.

2. Brewster, J. and Dangerfield, J.A. (1985), Chalk fields along the Lindenes Ridge, Eldfisk. Marine and Petroleum Geology 1 (3), 239-278.

3. Brooks, J. and Glennie, K.W. (Eds) (1987), Petroleum Geology of North West Europe.

4. Cameron, T.D.J., Stoker, M.S. and Lang, D. (1987), The history of Quaternary sedimentation in the U.K. sector of the North Sea basin. J. Geol. Soc. London 144, 43-58.

5. Caston, V.N.D. (1977), Quaternary Deposits of the Central North Sea. The Quaternary deposits of the Forties Field, northern North Sea. Rep. Inst. Geol. Sci. London 77/11, 21 p.

6. Caston, V.N.D. (1979), A New Isopachyte Map of the Quaternary of the North Sea. In. Oele, E., Schuttenhelm, R.T.E. and Wiggers, A.J. (Eds). The Quaternary History of the North Sea. Acta Universitatis Upsaliensis Uppsala, 23-8.

7. Curry, D., Adams, G.C., Boulter, M.C., Dilley, F.C., Eames, F.E., Funnel, B.M. and Wells, M.K. (1978), A correlation of Tertiary rocks in the British Isles. Spec. Rep. Geol. Soc. London 12, 72p.

8. Deegan and Skull (1977), a proposed lithostratigraphic nomenclature for the central and northern North Sea. Rep. Inst. Geol. Sci. London 77/25; Bull. Norw. Petrol. Direct. 1, 36pp.

9. Hardman, (1982), Chalk reservoirs of the North Sea.

10. Harland, R., Gregory, D.M., Hughes M.J. and Wilkinson I.P. (1978), A late Quaternary bio- and climatostratigraphy for marine sediments in the North Central North Sea. Boreas 7, 91-96.

11. Hovland, M. and Sommerville, J.H. (1985), Characteristics of two natural gas seepages in the North Sea. Marine and Petroleum Geology 2, 319-326.

12. Illing, L.V. and Hobson, G.D. (Eds) (1981), Petroleum Geology of the Continental Shelf of North West Europe. Heydon and Son, London, 521p.

13. Isaksen, D., Helliksen, D., Johansen, A. (1987), Top Layer Lithology/Pressures in Shallow Gas Resevoirs, in "A Shallow Gas Seminar", Norwegian Petroleum Directorate.

14. Jenyon, M.K. (1988), Overburden deformation related to the pre-piercement development of salt structures in the North Sea. J. Geol. Soc., London 145, 445-454.

15. Jenyon, M.K. (1988), Some deformation effects in an elastic overburden resulting from salt mobility, J. Pet. Geol., 11 (3), pp.309-324.

16. McAuliffe, C.C. (1979), Oil and Gas migration, chemical and physical constraints. Bull. Am. Assoc. Petrol. Geol. 63, 761-81.

17. Mitchell, G.F., Penney, L.F., Shotton, F.W. and West, R.G. (1973), A correlation of Quaternary deposits in the British Isles. Spec. Rep. Geol. Soc. London 4, 99p.

18. Nordberg, H.E. (1981), Seismic hydrocarbon indicators in the North Sea. Norwegian Symposium on Exploration, Bergen. Norwegian Petroleum Society, 8.1-8.40. (Gas chimney migration.)

19. Selley, R.C. (1978), Porosity gradients in the North Sea oil-bearing sandstones. J. Geol. Soc. London 135, 119-132.

20. Tissot and Welte (1984), Petroleum Formation and Occurrence. 2nd. Ed.

21. Woodland, A.W. (Ed) (1975), Petroleum and the continental shelf of North West Europe, Vol. 1: Geology, App. Sci. Publ., Barking, 501p.

PHILOSOPHY OF ADVANCED INTERPRETATIONS

B.P. Meier

Shell Internationale Petroleum Maatschappij B.V.

ABSTRACT

The interpretation of high resolution seismic plays a crucial role in shallow gas operating procedures. The results and reliability of the shallow gas investigation however rely on the input to the seismic interpretation. In practice, this input varies, depending on the amount of additional seismic and geological information which is consulted to support the interpretation, and depending on the method of computer-based techniques which are employed to enhance quantitative aspects of the seismic data analysis.
The standard approach in shallow gas data interpretation is considered to include the use of exploration seismic and well data, and the use of interpretation programs that offer the facilities to interactively digitise, depth-convert and contour seismic horizon data. This standard approach should be complemented by employing more advanced interpretation tools. A primary requirement are interactive workstations which thereby are substantially enhancing the efficiency and accuracy of the seismic analysis by providing the ability to interactively interpret the data and measure seismic horizon attributes.
Our aim to improve the reliability of the shallow gas prediction can only be achieved by choosing an integrated approach which uses modern seismic processing and interpretation techniques, which is embedded in a seismic and geologic frame provided by exploration data. This will ultimately contribute to minimising the risks associated with shallow gas in offshore drilling.

INTRODUCTION

The risks and consequences of shallow gas blowouts that have occurred in the past during offshore drilling have increased the oil industry's awareness of the associated hazards, and have resulted in considerable efforts to implement and improve shallow gas operating procedures. Accidents which have occurred more recently have further underlined the necessity and importance of carefully considering methods of high resolution seismic acquisition, processing and interpretation in order to come to a comprehensive and eventually reliable assessment of the shallow gas drilling hazards.

The outcome of the high resolution seismic interpretation relies heavily on the data quality which is, apart from the acquisition conditions, governed by a careful selection of the acquisition parameters and processing steps. Operational and economic constraints frequently dictate a simplistic approach which may considerably influence seismic data quality. With respect to safety in offshore drilling, this is certainly not an ideal situation. Our aim is to make the shallow

Volume 25: Safety in Offshore Drilling, 181–191.

gas prediction more reliable, and one of the advanced tools to persue this goal are interactive computer systems, which enable the analysis of reflection attributes.

The level of seismic interpretation varies considerably according to the methods and procedures applied. The shallow gas data usually cover a small area (e.g. 2km x 2km), and the record length is mostly confined to less than 1.5 seconds two-way-traveltime. This generally allows only for a limited insight into the geological domains where processes of potentially hazardous gas generation, migration and accumulation are taking place. The shallow gas interpretation can be considerably substantiated by taking account of available exploration seismic and well data which provide additional information regarding the probability of encountering hazardous shallow gas.

Ideally, synthetic seismograms should be used to compare seismic reflection data with well log data. Comparison of the reflectivity or acoustic impedance derived from seismic and wireline logs enables positive identification of horizons and matching of seismic events with geological features and potential shallow gas occurrences. However, in little explored areas this is rarely possible, and the interpretation has to rely on individual cases which under similar geological circumstances may be applicable. Even in more extensively explored areas, often little attention is given during logging to the shallow, top hole part of the stratigraphic section, such that the available data are generally unsuitable for the construction of synthetic seismograms.

The application of advanced seismic processing methods and interpretation tools for high resolution data, e.g. the use of interactive workstations and AVO processing, presently still requires an assessment of their practicality on a case-by-case basis. It will however become necessary to consider their routine use in order to carry out more efficient and thorough analyses of the shallow gas data.

SHALLOW GAS ASPECTS

'Shallow Gas'

Shallow gas in the context of safety in offshore drilling is defined as gas encountered in a well that cannot be closed in, because well shut-in pressures together with the hydrostatic head of the drill fluid in the borehole will probably result in formation breakdown and subsequent cratering of the well. If however BOP protection and normal well killing procedures can be applied, the gas is not considered to be 'shallow gas'. The term 'shallow gas' may therefore be applicable for gas pockets/reservoirs close to the surface as well as at greater depths.

Shallow Gas and Amplitude Anomalies

Our main concern during interpretation is the recognition of seismic features related to shallow gas and their assessment as potential drilling hazards within a given structural and stratigraphic framework. Shallow gas in the context given above has many aspects, and is expressed on seismic in various ways, depending on the mechanisms which form shallow gas accumulations.

Gas is often encountered in sedimentary formations with high porosity and high permeability. Depositions vary in thickness and lateral extent and may contain gas reservoirs which could be overpressured relative to the normal hydrostatic regime.

The seismic expression of shallow gas is dependent on the amount of saturation and its effect on the acoustic impedance contrasts of a given reservoir/seal sequence or of gas-saturated non-reservoir sediments. In the following, a few common examples of 'amplitude anomalies' are described.

Bright Spots: Shallow gas can be directly indicated by amplitude anomalies that either show a considerable increase in reflection amplitude, or a strong decrease resulting in a reflection phase reversal (Fig. 1). Transitional situations (dim spots) may also be expressed.

Pull-down or Sag: Acoustic travel time through a reservoir (or gas-saturated non-reservoir) is increased by the presence of gas. Arrivals from underlying reflections are delayed and they appear to be 'pulled down' or 'sag' as they pass underneath the gas zone (Fig. 2).

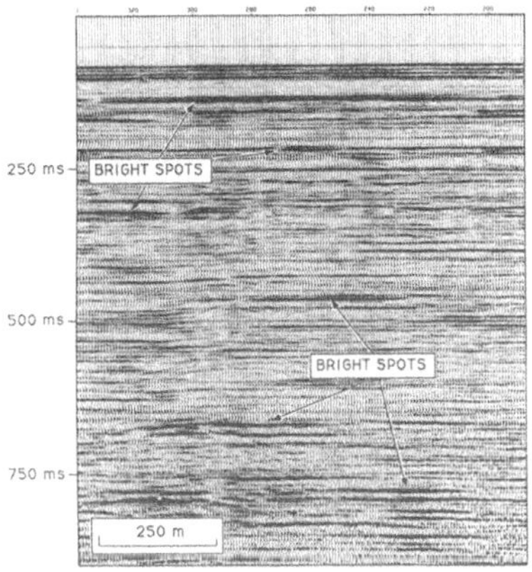

Figure 1: Examples of bright spots showing positive (black) and negative (phase-reversed) amplitude anomalies.

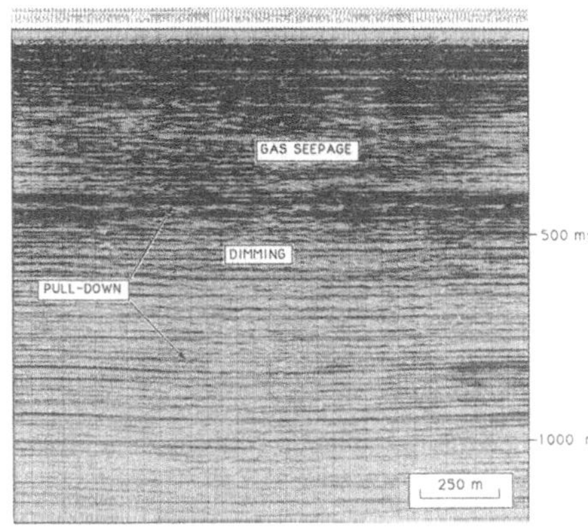

Figure 2: Zone of gas seepage above amplitude anomaly at ca. 400 milli-seconds. Note dimming and pull-down of reflections underneath.

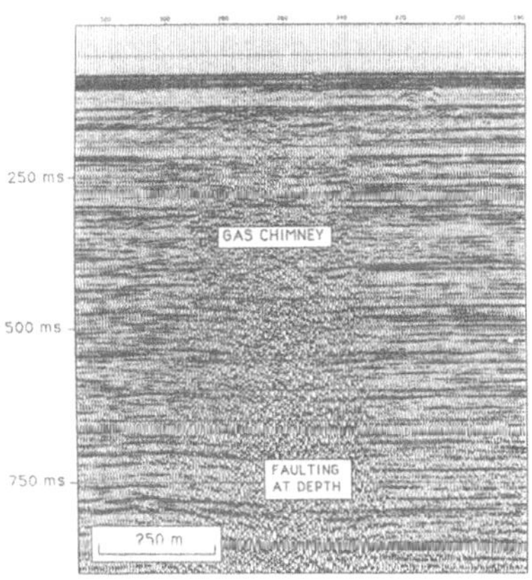

Figure 3: Example of a gas chimney. Bright spots are associated with the chimney at various levels.

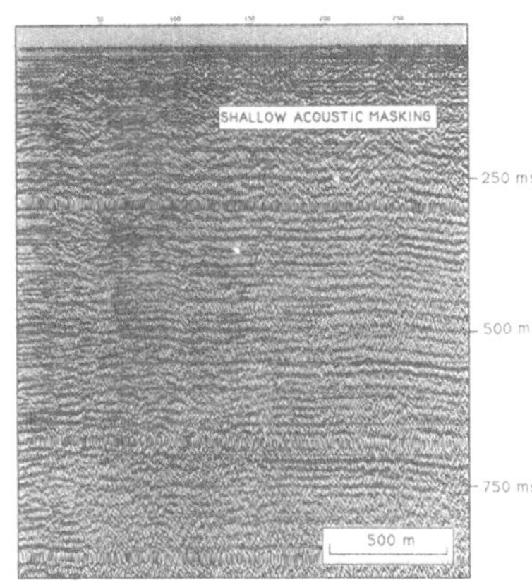

Figure 4: Acoustic masking of the deep section caused by organics in the near-seabed sequence.

Dimming: High reflection coefficients and absorption at the gas level may have a shadowing effect on the underlying sequence that is expressed by an overall dimming of reflections (Fig. 2).

Chimney: If gas from a deeper reservoir percolates up into the overlying sedimentary sequence, it can, although only partially saturating the sediments, have a significant effect on the seismic expression. This is shown by a reduction of signal strengths and zones of seismic incoherence which are called chimneys (Fig. 3). Individual or numerous amplitude anomalies and associated diffractions may also be recognised within or close to these gas seepage zones. As an entity they may also cause 'pull down' and dimming of underlying reflections. Gas chimneys are in some cases very obvious whereas in others they may be much subtler.

Shallow Acoustic Masking: Organic matter and biogenic gas, generated by in-situ decomposition of organic material, may have accumulated in the near-seabed sequence. The seismic response to these very shallow organic accumulations can be strong, causing severe acoustic masking of the underlying sequences (Fig. 4).

INTEGRATED SHALLOW GAS DATA INTERPRETATION

Shallow gas surveys acquired and processed to define possible accumulations of shallow gas in exploration areas have in recent years become commonplace. The purpose of the efforts to optimise acquisition and processing parameters is to facilitate the final and most critical stage of the shallow gas investigation which is the reduction of the data to a realistic model of the subsurface strata, predicting potentially hazardous shallow gas occurrences and illustrating both structural and geologic characteristics.

Improvements in data acquisition and processing techniques have significantly increased the informational content of seismic data, and the improved quality often provides additional evidence for depositional history and lithology. It is therefore important that processing and interpretation are coordinated at an early stage.

Needless to say, the results and reliability of the shallow gas investigation strongly depend on the input to the seismic interpretation. In practice, the approach in high resolution data interpretation may however vary, depending on the amount of additional information considered to support the interpretation, and on the computer-based techniques employed to enhance quantitative aspects of the seismic data analysis.

Basic and Standard Elements

The input to the seismic interpretation is frequently limited to the absolute minimum, i.e. the seismic sections with the corresponding shot point map, and little or no additional seismic or geological information. This basic approach, although often seemingly enforced by operational constraints, is certainly ill advised and conceptually unacceptable. On the other hand, taking account of additional information that is available, and assisting the interpretation with computer-based techniques, may improve the shallow gas prediction but does not guarantee it. The basic and standard elements, which are discussed in the following, are however considered as fundamental requirements for an integrated approach in the shallow gas data interpretation.

High Resolution Seismic Data

Although no set rules can be used to determine which parameters are optimal for a certain dataset, some general comments are worth making. The ultimate goal of the seismic processing is to obtain a reliable representation of the subsurface geological successions by removal from the seismic record of non-geological information such as noise and seismic artefacts. On the other hand, our goal is also to retain as much primary seismic information as possible. This approach, of course, requires a careful balancing of the processing parameters on a case-by-case basis.

From the interpretation point of view, amplitude, phase and frequency criteria play an important role, not only in the assessment of amplitude anomalies, but also in the recognition of (seismic-) stratigraphic relationships. Amplitude level correction is therefore a critical task which needs to be borne in mind at the corresponding stages of processing such as gain recovery, trace equalisation and amplitude equalisation (AGC). Similarly, fixed or time variant bandpass filtering have also to be carefully considered.

Usually two sets of seismic displays are required: the equalised stack and relative amplitude. The equalised stack display, to which a final sliding window gain is applied, provides a section for 'structural' interpretation, whereas the relative amplitude display is suitable for the interpretation of amplitude variation in terms of stratigraphy. Occasionally, displays of equalised stack sections -with a wide AGC window applied- can also be used for a basic assessment of amplitude variations.

The vertical and horizontal scaling of a seismic section can significantly influence the interpretation. Trace spacing, if too tight, can deteriorate data quality because the amplitude information will be limited by the dynamic range of the plot, and can therefore not be fully displayed. A similar loss in clarity may be incurred if inadequate vertical scales are chosen.

Exploration Data

Generally, high resolution seismic data and possible amplitude anomalies as potential drilling hazards cannot comprehensively be interpreted without taking account of existing exploration data. Exploration seismic and geological data provide the necessary framework for the shallow gas seismic.

(a) Regional Scale

The seismic data on a regional scale provide valuable information about the overall depositional setting for the stratigraphic sequence. In many offshore areas, the 'shallow' sequence (ca. 1-2 kilometers thick) is frequently composed of young, Tertiary-Quaternary basin fill sediments which were deposited under the influence of active deltas. Reflection configurations on regional seismic data may outline the structure of the delta which according to a classical concept is formed by sand-prone topset, predominantly silt- and clay-prone foreset and mainly clay-prone bottomset deposits (Fig. 5). These three major sub-divisions provide a first indication for the overall sediment composition, and consequently for the probability of potential reservoir and seal deposits being present in the section displayed on the high resolution seismic.

Regional seismic data also provide useful information about the structural regime of an area within which the site survey data was acquired. Faults visible on the high resolution seismic may, for example, correspond to the shallow part of deep-seated structures (Fig. 6). On the other hand, if the area is affected by synsedimentary tectonics, unconformities may be recognised more easily on the regional seismic, and a better distinction between depositionally and structurally controlled variations in sediment thicknesses can be made.

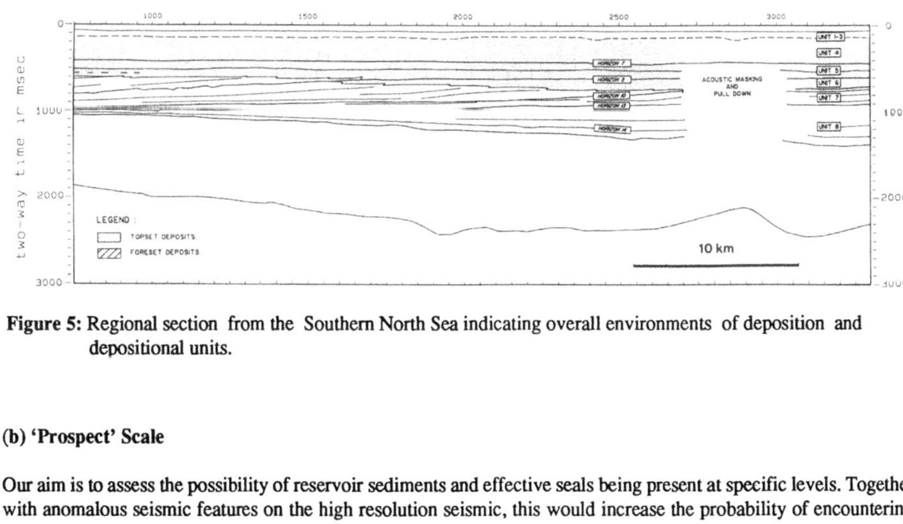

Figure 5: Regional section from the Southern North Sea indicating overall environments of deposition and depositional units.

(b) 'Prospect' Scale

Our aim is to assess the possibility of reservoir sediments and effective seals being present at specific levels. Together with anomalous seismic features on the high resolution seismic, this would increase the probability of encountering shallow gas.

Exploration seismic in a stricter sense, i.e. covering only the prospect area, assists in defining stratigraphically relevant depositional sequences and units in the shallow part of the section by applying principles of seismic-stratigraphic interpretation (Vail et al., 1977). The analysis of their lateral and vertical relationships may help to detail the overall environments of deposition as they were recognised on a regional scale or on the prospect scale itself. The exploration seismic data also helps to define fault trends and patterns in the shallow section which may become relevant as potential migration paths for gas generated at greater depths. In addition, faults outside the site survey area, which may form part of a combined dip/fault closure in the shallow part of the section, and the lateral extension of eventually known reservoirs beyond the high resolution data coverage can be pin-pointed (Fig. 6).

Long high resolution seismic lines, i.e. with a length similar to the exploration seismic lines, can be very useful in further detailing seismic and geological features recognised on a prospect scale.

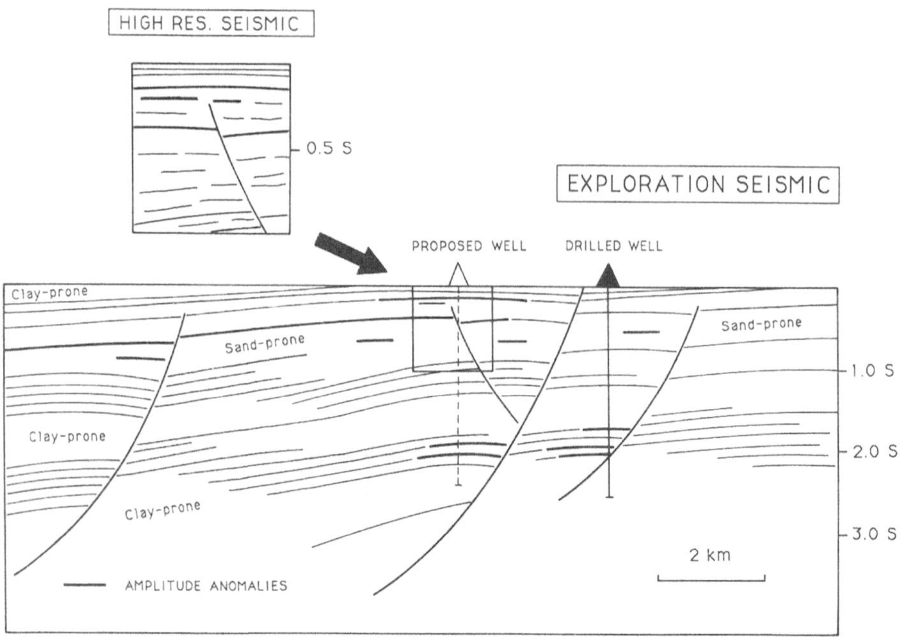

Figure 6: Exploration seismic and well data provide the frame for the high resolution seismic, and help to clarify the nature of amplitude anomalies expressed on the shallow gas data.

(c) Well Data

For the positive identification of lithofacies as established from the seismic data a calibration well is required. In the absence of such a well, subjective judgement based on comparable, previously calibrated examples from the same or close-by areas can be made.

Anomalously high amplitude reflections originate from strong impedance contrasts in the subsurface which may or may not be caused by gas layers. Organic sediments with a low acoustic impedance such as peat, lignite or coal can produce a similar effect on seismic, and a brine with a gas saturation as low as 5% will reduce acoustic velocity to cause an amplitude anomaly. On the other hand, carbonate layers consolidated by early diagenesis may also give rise to very high amplitudes. Tophole well data or borehole data help to identify such layers, and eventually to distinguish between amplitude anomalies which are caused by gas or strong lithological contrasts.

Geological Models

An important aspect of high resolution data interpretation is the assessment of the potential for shallow gas generation. Biogenic gas may be generated in rapidly accumulating marine sediments by bacteria living in anoxic and sulphate deficient environments with relatively low temperatures ranging from 0 - ca. 75 degrees Celcius (Barker, 1987; Rice and Claypool, 1981). A minimum amount of metabolisable organic matter has to be present, and burial depths, typically, should not exceed 2000 metres. In fact, the timing of the factors that control biogenic (methane) gas generation is such that most gas is generated prior to burial depths of 1000 metres, after having reached its peak at a depth range between 300-400 metres.

Geological models based on seismic and well data describing environments of deposition, depositional history and sediment properties will eventually provide valuable information with regard to the potential of shallow gas generation, migration and accumulation. Knowledge of the shallow sedimentary sequence as detailed as possible, together with the structural interpretation, will enable a realistic assessment of possible migration paths for gas generated at greater depths as well as at shallow depths. Migration paths may be either 'diffuse', e.g. through inter-connected, stacked channel sequences, or 'discrete' along fault planes and laterally along slightly tilted sand/clay alternations.

Computer Assisted Interpretation

A standard requirement for the structural analysis of the shallow gas seismic data is the use of an interactive interpretation package that runs on a desktop computer, preferably linked to a main frame computer for more advanced applications, and handles input and output of topographic files on diskette or via a main frame device.

The package should offer the facilities to interactively digitise interpreted seismic sections, and to time-depth convert the digitised horizon data with a suitable conversion method such as linear time and depth velocity functions, time-depth table velocity functions (2-way tables), time/velocity tables or a user defined velocity function. The system should also be capable of gridding and contouring horizon data, of creating time and depth contour and feature maps, of handling fault features and of plotting the data at appropriate scales.

The use of an interactive system makes the structural analysis and mapping of shallow gas related seismic features significantly more effective and accurate, and a realistic time-depth conversion greatly assists in establishing the drilling programme and planning of drilling operations.

<u>Advanced Elements</u>

In addition to the standard elements discussed above, the use of interactive workstations and, if sufficient top hole well data is available, the construction of synthetic seismograms might be envisaged, taking thus full advantage of modern, present day interpretation tools. Both tools are indispensable elements working towards a quantification of the high resolution seismic data.

The use of interactive workstations is primarily recommended for the interpretation of large datasets, and for the analysis of high resolution seismic from known, gas-prone areas where stratigraphic and structural evidence, together with anomalous seismic features strongly suggest the presence of shallow gas.

Interactive Workstations

In recent years, interactive workstations have become widely used for the interpretation of 3D and 2D seismic data. Their use has greatly facilitated seismic analysis by speeding up the laborious work of horizon mapping, and has considerably improved the interpretative work by enabling a measurement and display of various aspects of reflection data such as relative amplitude, instantaneous phase and instantaneous frequency. On seismic data in general, increased reflection amplitudes, phase reversals, high frequency loss and low velocities are seismic attributes which can be indicative of shallow gas accumulations. The benefit of being able to detect variations in these attributes on the high resolution seismic data is obvious.

Measuring horizon attributes requires reliable automatic tracking programs. Certain advantages over manual line by line picking follow, most importantly the enhanced phase consistency of the picks throughout the grid. This presents the opportunity to calculate horizon attributes sensitive to geological and potentially shallow gas related variations.

The colour display of amplitude values on the interactive workstation exceeds the dynamic range of conventional paper displays, and may provide valuable information about the presence of shallow gas and its migration path (Fig. 7). The ability to extract (relative) amplitude values and amplitude mapping of the event of interest highlight areas defined by phase reversals and/or areas with higher than normal values (Fig. 8). Combined with structural data from the same level, this additional amplitude information can be used to further assess potential shallow gas occurrences.

The option to calculate complex seismic trace attributes such as instantaneous phase assists in the interpretation of reflection configurations and terminations in zones of low or strongly variable reflectivity since all amplitude information is discarded. Instantaneous phase displays can be useful to distinguish between 'genuine' amplitude anomalies and 'apparent' amplitude anomalies caused by tuning of reflections ('loop-mixing'), to pin-point the extent of amplitude anomalies, and, generally, to better assess continuities of reflections with laterally varying amplitudes in low amplitude areas.

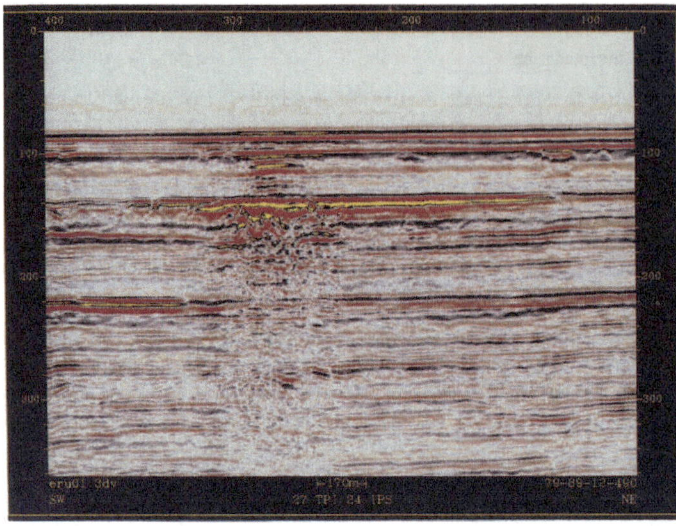

Figure 7: Colour display of a high resolution seismic section on an interactive workstation. Negative amplitudes are shown in red, positive amplitudes in black, and high positive amplitudes in yellow. Note the increased amount of amplitude information in the near-seabed sequence (above ca. 130 ca. milli-seconds) overlying a gas chimney (compare with Fig. 3).

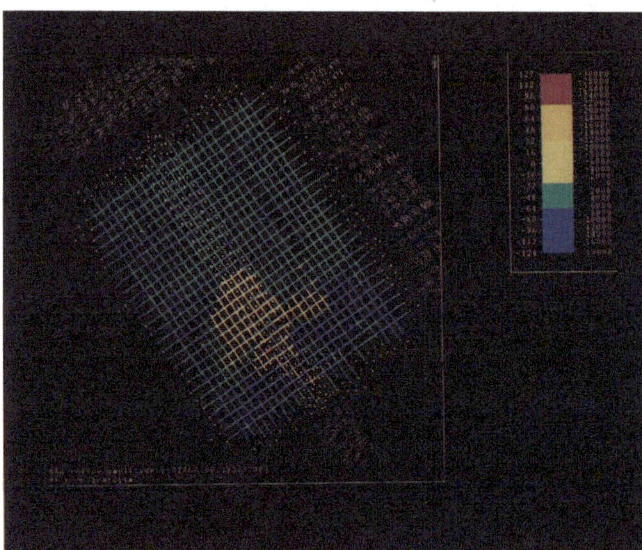

Figure 8: Amplitude map of a horizon from a 2D, high resolution data set. The yellow and red colours highlight areas with reflection phase reversals. Note the limitations to establish the areal distribution tion of potential shallow gas in some places, caused by the 2D nature of the seismic.

The use of instantaneous frequency or average weighted frequency displays may bring out the shadow effect of high frequency loss underneath a potential shallow gas accumulation which is caused by absorption in the porous gas layer.

Synthetic Seismograms

Wherever available or feasible, the use of synthetic seismograms should be envisaged to compare the shallow gas data with well log data. The comparison of the reflectivity, or acoustic impedance, derived from seismic and wireline logs enables positive identification of horizons and matching of seismic events with geological markers. In some cases it might be possible to relate certain seismic attributes to porosity and porefill. Variations of these attributes away from the borehole are then indicative for changes in the established relationship.

In order to make the well data comparable with seismic data, the acoustic impedance derived from wireline logs has to be converted from depth to time, creating thus the reflectivity. It is then necessary to resample the data to an increment which is comparable to that of high resolution seismic. The high resolution seismic data, on the other hand, has to be further processed to zero-phased, acoustic impedance sections.

Once the synthetic seismograms are prepared, it should be possible to transfer them to the processing environment of the seismic data which will enable the seismic-to-well match by splicing the synthetics into the seismic section. This will eventually also enable the measurement of seismic attributes within the seismic processing environment, and the use of the synthetics on an interactive workstation.

Integrated Interpretation and Reporting

The results obtained by applying standard and advanced interpretation techniques have to be synthesized into a comprehensive shallow gas assessment, by concisely evaluating the contributions of the various elements. The objective of the shallow gas survey is to identify, assess and map possible gas accumulations. It is essential that the results of the shallow gas interpretation find their way into the well proposals and drilling programmes, to ensure that the necessary steps and drilling procedures are adopted.

The shallow gas report must address geophysical as well as geological aspects of the interpretation by briefly describing the equipment used, the processing and quality of the data, and the structural geology, stratigraphy and depositional history of the area. The report should contain a dedicated section on amplitude anomalies and their assessment as potential drilling hazards, clearly indicating the depth and lateral extent of any possible accumulations. Careful attention should be given in defining the origin and potential effects of shallow gas anomalies by considering the likely geologic and structural history and, by inference, whether the anomalies are capable of overpressuring and thus being a potential problem.

Charting should not only be limited to maps showing the depth and lateral extent of potential gas accumulations but also include depth contour maps at corresponding levels, and of any relevant geological reflectors which enable an evaluation of shallow gas trapping mechanisms. Using interactive workstations, the amplitude anomaly and structural mapping is replaced and/or complemented by horizon attribute maps at the corresponding depths.

In the event that shallow gas anomalies are present in the data, the report should recommend certain possible courses of action. The simplest and safest course of action is to propose a drilling location such that the well will not penetrate potential shallow gas accumulations. It may however not always be possible to avoid drilling through potential shallow gas layers without having an impact on the deep targets. Consequently, the decision may be taken to drill the well knowing that anomalies are present at clearly defined and highlighted levels. It is essential in this case that specialist staff of concerned parties come together to fully consider the necessary steps and safety procedures exhaustively to which a comprehensive, integrated shallow gas interpretation forms the basis.

ENHANCEMENT OF SHALLOW GAS DATA INTERPRTETATION

With regard to digital high resolution seismic acquisition and processing, it is evident that the present approach to shallow gas prediction has not yet taken full advantage of available, modern techniques which enable a further enhancement of the gas hazard assessment in the tophole section. In order to improve quantitative aspects of the shallow gas data interpretation, it is however inevitable to persue the application, and examine the feasibility and practicality of these techniques.

In terms of **acquisition**, it would appear that 2D acquisition is approaching a technical plateau, and it will become necessary to investigate the usefullness of conducting small high resolution 3D surveys, even though it may presently have a significant financial impact on the total survey costs. The use of 3D seismic would undoubtedly enhance the potential to predict porefill and lithology as the amplitude information on 3D seismic is more reliable than on the 2D data.

In areas of field development and in known, gas-prone areas **logging** of the tophole part of the stratigraphic section can be further improved by running the required sonic and density logs for the construction of synthetic seismograms. Matching the seismic to the well data will eventually result in a better understanding of the seismic and lithological characteristics of a particular area, which in turn allows for a better assessment of the parameters relevant for shallow gas generation and trapping.

An analysis of **AVO** (Amplitude Versus Offset) effects may assist in distinguishing gas related amplitude anomalies from other types of anomalies, provided there are no geometrical restrictions imposed by a limited offset. For a given porous lithologic unit, the behaviour of the P-wave reflection amplitudes versus offset will vary, depending on whether the pore spaces are filled with gas or water. A gas bearing sand layer may reveal a marked increase in reflection amplitude with offset, whereas a water bearing sand layer may show a small decrease. AVO processing aims to extract this amplitude behaviour -lost in conventional stacking- from pre-stack data. Therefore, careful pre-stack processing is very important to preserve amplitudes as much as possible, and to obtain pre-stack data with high signal-to-noise ratios.

It is essential to consolidate the approach of using **interactive workstations** for the interpretation of shallow gas datasets. This will result in a critical assessment of their best use for the future requirements, and pave the way for integrating further, advanced, software-driven seismic interpretation techniques such as a quantified seismic reflectivity analysis which results from processing of reflection attributes on 2D seismic data.

An important aspect of shallow gas data interpretation is establishing and maintaining a longterm **database** which is continuously updated with results of shallow gas investigations, and calibrations of high resolution seismic datasets. This will eventually enable an assessment of the shallow gas risks in exploration and field development areas, which in turn can be used as input to future, integrated site survey studies.

CONCLUSIONS

Our aim to make the shallow gas prediction more reliable can only be achieved by choosing an integrated approach which uses modern, up-to-date processing and interpretation techniques applied to high resolution seismic datasets. This approach should be embedded in a seismic and geologic frame provided by exploration data.

It is indispensable that the standard approach, which takes account of exploration seismic and well data, is complemented by the use of advanced interpretation tools to improve quantitative aspects of shallow gas data interpretation. A primary requirement is the routine use of interactive workstations, which substantially enhance the efficiency of seismic analysis, and offer the ability to measure, calculate and map reflection attributes that are sensitive to geological and shallow gas-related variations. Gaining further experience in the application of interactive workstations will result in maximising their use with regard to the interpretation requirements, and facilitate the integration of additional advanced seismic processing and interpretation techniques.

The present approach in shallow gas data interpretation will in the future undoubtedly benefit from the more extensive use of further methods to enhance the shallow gas prediction, such as 3D seismic and AVO studies. This will ultimately increase safety in offshore drilling by minimising the risks which are associated with shallow gas.

ACKNOWLEDGEMENTS

This paper has benefited from numerous internal reports and discussions with colleagues. The writer appreciates Shell Internationale Petroleum Maatschappij's permission to publish the paper.

REFERENCES

Barker, C. (1987). Development of abnormal and subnormal pressures in reservoirs containing bacterially generated gas, AAPG Bull., Vol. 71, 1404-1413.

Rice, D. and Claypool, G. (1981). Generation, accumulation, and resource potential of biogenic gas, AAPG Bull., Vol. 65, 5-25.

Vail, P.R., Mitchum, Jr., R.M., Todd, R.G., Widmier, J.M., Thompson III, S., Sangree, J.B., Bubb, J.N. and Hatlelid, W.G. (1977). Application of seismic reflection configuration to stratigraphic interpretation. In AAPG Mem., No. 26, 47-212.

'SIGNATURE' THE FUTURE APPROACH TO RELIABLE SHALLOW GAS DETECTION

J.P. Williams and M. Sarginson

Hydrosearch Associates Ltd, Chandler House, Knaphill, Woking, Surrey, UK

INTRODUCTION

In May 1988 Hydrosearch commenced work on the 'SIGNATURE' project which concentrated on the usage of a computer workstation for the detection of shallow gas at proposed drilling locations.

Phase I of the project, which concentrated on analysis of post-stack data, was part-funded by the Offshore Supplies Office of the UK Department of Energy with the remaining funding being provided from the Hydrosearch R & D budget.

Volume 25: Safety in Offshore Drilling, 193–208.

Phase II of the development programme involved further work on stacked data together with some analysis of pre-stack data. This phase of the project was supported by two major international oil companies.

Some key aspects of the work undertaken during Phase I and II of the 'SIGNATURE' programme are summarised in the following figures and commentary.

The authors regret that time constraints have prevented a full manuscript from being prepared.

'SIGNATURE' PROJECT
- OBJECTIVES -

- To provide proven diagnostic procedures for Shallow Gas detection
- To minimise subjectivity and improve consistency in interpretation
- To improve reliability of prediction of drilling hazards

The above must be achieved within normal site survey timescales

FIGURE 1 HYDROSEARCH

The overall objectives of the 'SIGNATURE' development programme are shown in Figure 1.

The diagnostic procedures include analysis of both pre- and post-stack data and require a standard methodology to be followed and optimum display parameters (scales, colour palettes, etc) to be utilised.

The objective has been to develop an approach which can provide a detailed analysis of up to four lines from a shallow gas survey grid within 72 hours of receipt of the SEG Y stacked data.

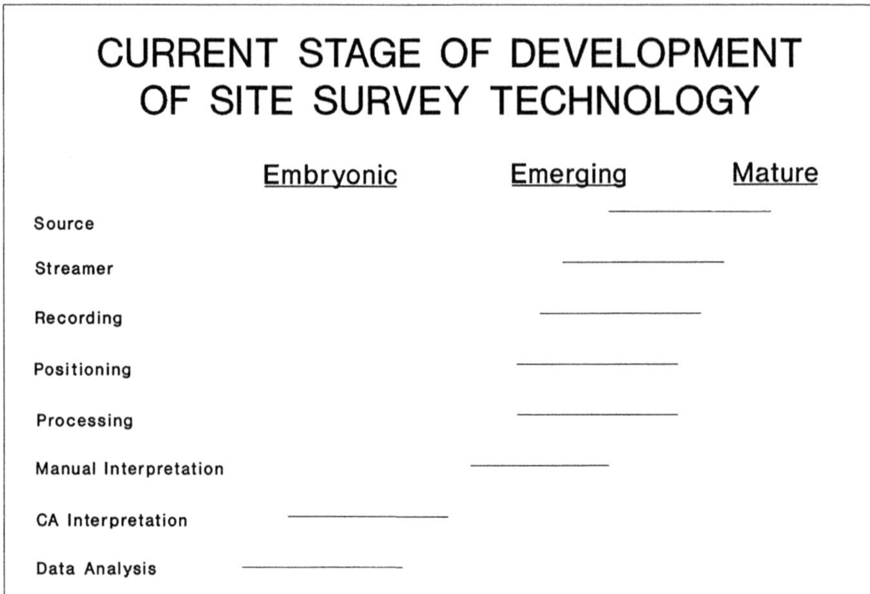

FIGURE 2

Before describing the 'SIGNATURE' developments to date it is worth looking at the relative stages of development of the key components of shallow gas survey technology.

This slide shows quite clearly that the key acquisition components are at a relatively mature stage in their development although there is still considerable scope for new methods of application (e.g. 'engineering' 3D).

It is in the areas of Computer-Aided (CA) Interpretation and Data Analysis that development effort now needs to be concentrated if shallow gas detection is to improve. The enabling mechanism for this is the use of workstations for interpretation and analysis of gas hazard data.

It is interesting to note that the engineering geophysicist is lagging behind his exploration colleague by about five years in the routine use of such tools.

POST-STACK DIAGNOSTIC CRITERIA

- AT THE LEVEL OF THE ANOMALY
  - Amplitude/Reflection Strength
  - Instantaneous Phase
- BELOW THE LEVEL OF THE ANOMALY
  - Pull-down
  - Mis-stacking
  - Amplitude Blanking
  - Frequency Anomalies

FIGURE 3 HYDROSEARCH

Phase I of the 'SIGNATURE' development programme concentrated on the analysis of stacked data. The primary diagnostic criteria used to establish whether or not an anomaly is caused by the presence of gas are shown in Figure 3.

On the basis of the use of these criteria three main conclusions were drawn from the work undertaken during Phase I.

1. All the diagnostics must be tested for every data set and although different confidence weighting can be placed on the various criteria each one can contribute to the overall conclusions. The key issue is "use all the available information".

2. There are limitations in the level of discrimination which can be achieved by post-stack analysis eg. gas sand vs lignite.

3. If data are to be analysed on a workstation, acquisition and processing quality control must be rigorous.

VALUE OF POST-STACK DIAGNOSTICS

DIAGNOSTIC	COMMENT ON VALUE
Amplitude/Reflection Strength	Reliable diagnostic both in True Amplitude and Clipped form
Instantaneous Phase	Good for determing polarity reversal, recognising diffractions and tracking weak events
Pull-Down	Sensitive parameter for identifying velocity anomalies
Mis-Stacking	Sensitive parameter indicative of lateral velocity variations
Amplitude Blanking	As above plus absorption of energy from raypaths passing through a gas sand
Frequency Anomalies	Instantaneous and FFT methods generally unreliable
Attribute Extraction	Provides numerical data for mapping and statistical analysis
Attribute Mapping	Useful for relating anomaly locations to topography/geological features

FIGURE 4 HYDROSEARCH

The relative value of the post-stack diagnostics used during Phase I and II are summarised in Figure 4.

The comments are based on our experience in analysing approximately thirty gas hazard data sets from the UKCS and various areas offshore worldwide.

It has become apparent that the value of the individual diagnostic criteria will vary depending on the data being analysed but in all cases, amplitude and phase characteristics have proved valuable. Although not yet utilised routinely, attribute mapping is also seen as a key component, enabling a 'pseudo-3D' view of the data to be created.

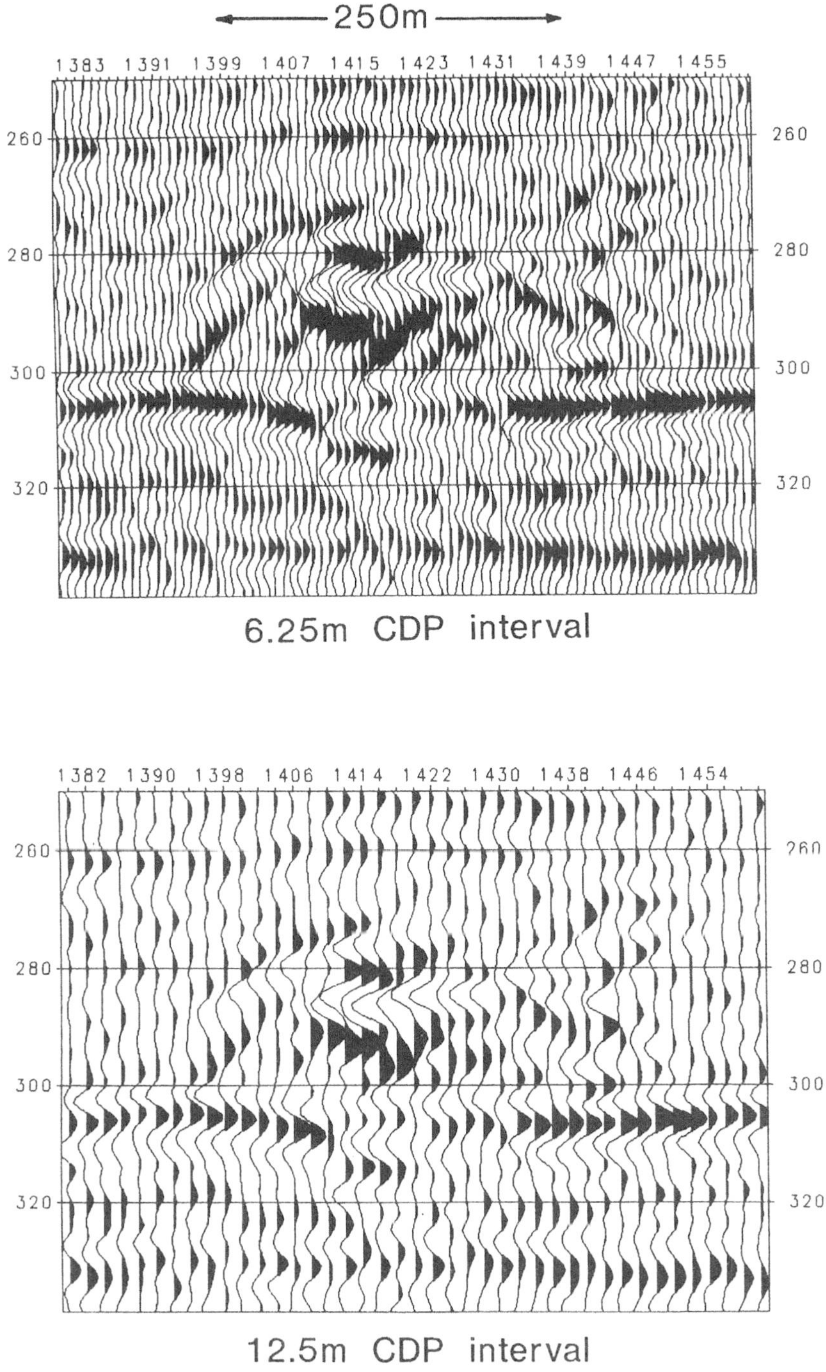
250m
1383 1391 1399 1407 1415 1423 1431 1439 1447 1455
260
280
300
320
6.25m CDP interval
1382 1390 1398 1406 1414 1422 1430 1438 1446 1454
260
280
300
320
12.5m CDP interval

FIGURE 5

One example of the value of display flexibility available with a workstation is shown in Figure 5.

This data example illustrates a mounded feature with high amplitude characteristics displayed at an exaggerated vertical scale of 100cm/second. The improvement in lateral resolution when using a 6.25m CDP display interval is clearly seen. In particular, the dipping events marking the flanks of the feature in the CDP range 1425 to 1445 are aliased on the 12.5m CDP display at the frequencies present in this data. The closer sampling on features with small lateral dimensions in the plane of the sections can also help in resolving amplitude variations which are a function of the Fresnel Zone dimensions.

This form of 'feedback' into acquisition parameters is invaluable and validates the theoretical calculations for spatial aliasing which will be used in planning cross-line spacing for any engineering 3Ds.

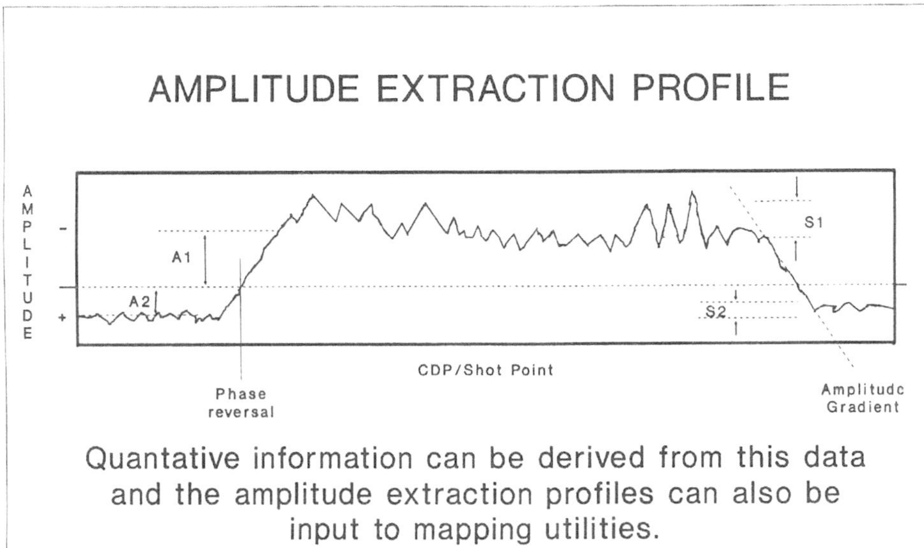

FIGURE 6

A recent addition to the 'SIGNATURE' system has been the development of an attribute extraction utility which enables instantaneous attributes to be extracted along any interactively digitised horizon.

To date, effort has concentrated on the extraction of amplitude profiles as shown in Figure 6. This information can then be fed to an attribute mapping utility to produce horizon attribute maps and it can also be analysed for amplitude ratios and statistical measures of amplitude variation within specific CDP ranges.

It is anticipated that this type of analysis, possibly performed in a semi-automatic manner, may assist in discriminating between gas-induced seismic anomalies and seismic responses from units such as lignite.

SIGNIFICANT PRE-STACK PARAMETERS

- AVO effects

- The trequency, location and accuracy of velocity analysis

Analysis of such data can be time-consuming and requires a high level of interaction between 'interpretation' and 'processing'

FIGURE 7 HYDROSEARCH

The limitations of 'post-stack only' analysis for the detection of shallow gas have already been noted. The Phase II development of 'SIGNATURE' included analysis of pre-stack parameters which could either assist in discriminating between anomalous features or alternatively could have a significant influence on the quality and reliability of the stacked data.

Amplitude versus Offset (AVO) falls into the first category, the frequency, location and accuracy of velocity analysis, into the latter.

By essentially bringing analysis forward into the pre-stack phase the processing/interpretation boundary becomes hazy and points towards an integrated approach for these phases in the future.

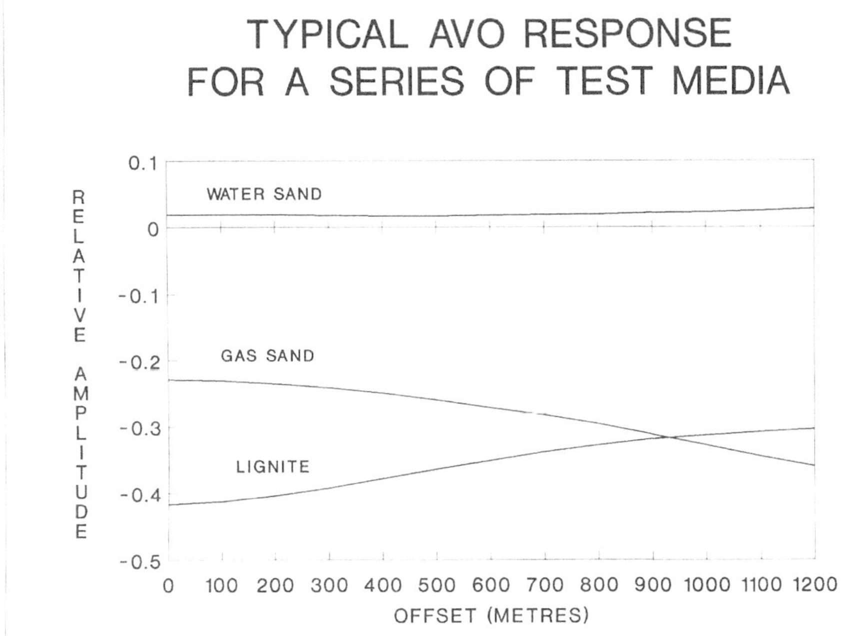

FIGURE 8

Initial work on Amplitude vs Offset has concentrated on forward modelling of AVO responses for typical North Sea features.

The response curves shown in Figure 8 have been produced using the following physical parameters - regarded as typical for a Central North Sea anomaly at approximately 550 ms two-way travel time.

Parameter	**Encasing Claystone**	**Water Sand**	**Gas Sand**	**Lignite**
Vp m/s	1750	1800	1200	1200
Vs m/s	714	735	788	362
Poisson's Ratio	0.40	0.40	0.12	0.45
Density g/cc	2.00	2.02	1.82	1.20

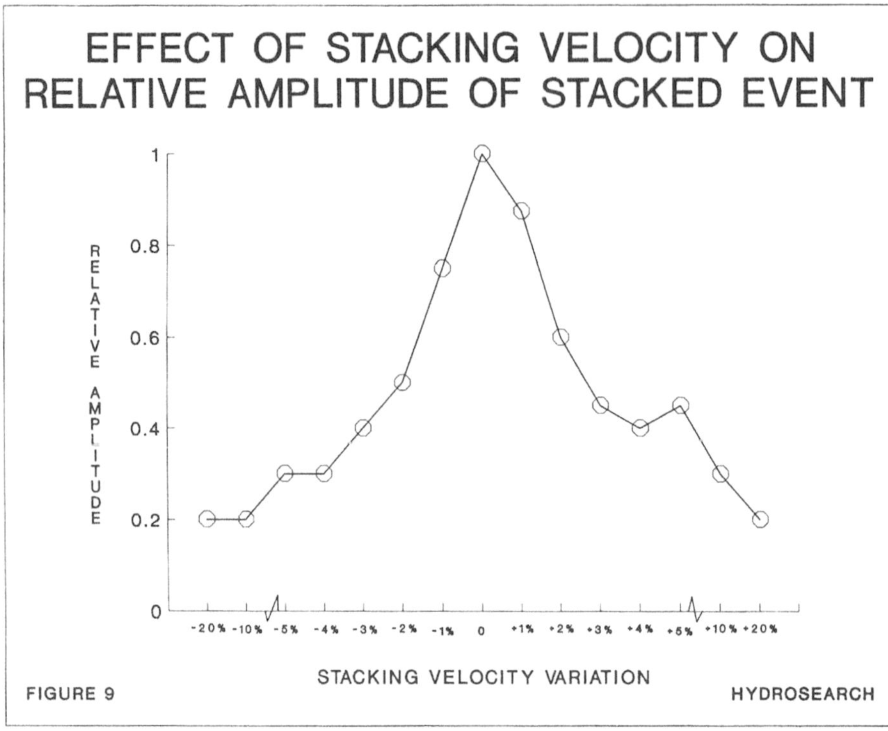

FIGURE 9

Sensitivity studies on these parameters have indicated that realistic variations in parameters will not alter the overall form of the response curves. Future effort in this area will concentrate on analysis of actual AVO data.

The key role that stacking velocity can play in determining the characteristics of stacked data has been studied in some detail. In particular, relative amplitude and frequency content in the resultant stack have been analysed as a function of variations in stacking velocity on actual North Sea data.

Figure 9 shows the amplitude variation which can be produced by 'errors' in stacking velocity. Note that for a typical event at 400ms TWT a deviation of only 2% from the optimum velocity can

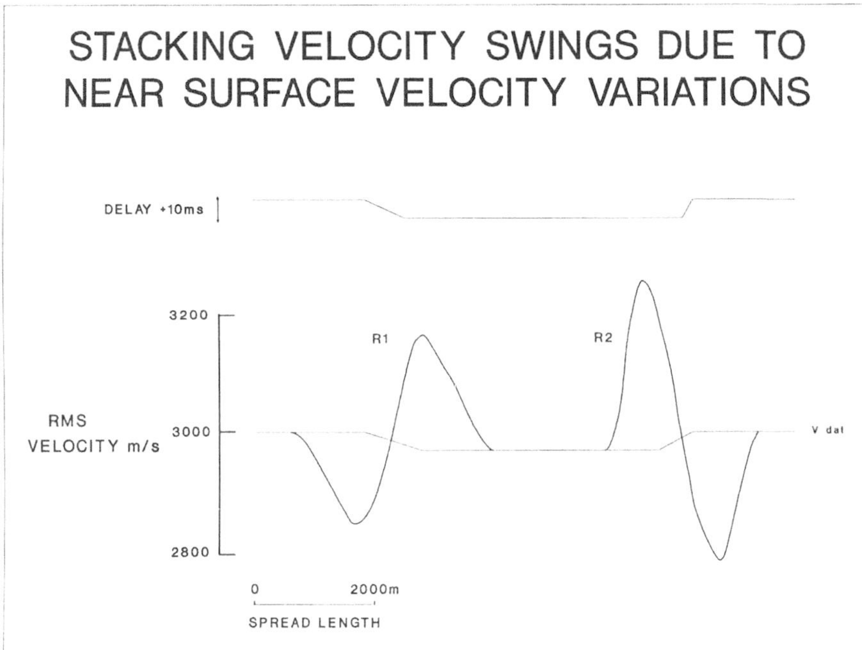

FIGURE 10* after Al-Chalabi, 1979.

reduce the amplitude of the stacked response by 50%. This has obvious implications if amplitudes are to be used in a quantitative manner to assist in identifying shallow gas.

Errors in stacking velocity can occur either from mis-picking or due to interpolation between analysis points. Figure 10 shows the type of stacking velocity variations which can occur when some offsets within a CDP gather are affected by a near-surface velocity anomaly. In the North Sea, this magnitude of raypath delay can be introduced by an infilled channel with a 'depth' of 400ms having a 6% velocity contrast which is typical of channel infill in Central North Sea.

KEY COMPONENTS FOR SUCCESSFUL PREDICTION OF GAS

- Data which has been acquired and processed correctly
- Detailed analysis and cross-correlation of available data and results
- Acceptable geological model
- Experience

FIGURE 11 HYDROSEARCH

The problem increases substantially in the presence of vertically co-incident gas anomalies, which by their very nature produce significant lateral velocity variations. Note that in the presence of such variations, no stacking velocity is 'correct' but merely a best estimate which maximises amplitude and minimises frequency loss of the stacked event.

The 'SIGNATURE' project has shown the need for careful quality control of data throughout the acquisition and processing phases if the data are to be analysed quantitatively on a workstation.

Analysis of data must include all the available information, a key component being the need to explain any conclusions within the framework of a plausible geological model.

Finally, experience in the interpretation of gas hazard survey data <u>and</u> in the use of workstations is essential if reliable results are to be produced.

FUTURE REQUIREMENTS

- Increased utilisation of the workstations and guidelines for their use for drilling hazard detection
- Better understanding of the limitations of stacked data
- Closer interaction between processing and interpretation phases
- Increased effort into analysis of data - feature analysis, pattern recognition, etc.

FIGURE 12 HYDROSEARCH

The benefits which can be gained from the use of workstations for the analysis of gas hazard data have been shown during the 'SIGNATURE' development programme. This trend must continue if confidence levels in predicting gas are to increase. It is felt that the generation of guidelines for the use of workstations for gas hazard analysis could assist interpreters significantly.

As interpretation becomes more detailed, the limitations of stacked data must be clearly understood such that the interpreter is aware of the limitations both of the techniques and the data at his or her disposal.

As a part of this, and in particular as pre-stack analysis increases, processing and interpretation need to become more unified.

Finally, as illustrated earlier, Computer Aided Interpretation and Data Analysis as applied to gas hazard data are still in their infancy. We would now expect to see an increase in the effort put into these areas with the objective of developing some semi-automatic aids to gas hazard detection over the next 2 to 4 years.

Part IV
Application of Geophysical Data

PREDICTION OF SHALLOW GAS FROM SEISMIC DATA

Reidar Kanestrøm, Åshild Skålnes, Per Riste, Trond Eide and Sverre Strandenes.

Norsk Hydro A/S, Research Centre, Bergen, Norway.

ABSTRACT

Since prediction of shallow gas is essentially a prediction of pore fill, we have chosed to focus on seismic parameters which are significantly affected by a change in pore fill. These parameters are: Seismic velocity, attenuation, amplitude versus offset and seismic response. To study these parameters, different techniques are required, and in this paper we apply the techniques on real data for shallow gas prediction. If all the techniques described could be applied at the same site, the result should be a reliable prediction.

The attenuation of seismic energy is a promising parameter, but it requires that refracted waves are recorded.

The amplitude versus offset is expected to be a powerful prediction technique in the future, but at the present time more development is needed. This technique also has the potential of estimating the saturation under favourable conditions.

Volume 25: Safety in Offshore Drilling, 211–232.

Single trace inversion is also a promising technique, and with further development and linked to velocity modelling it will become a powerful tool in shallow gas prediction.

Direct use of seismic velocities in shallow gas prediction requires refracted arrivals. These data are of special importance where no borehole data are available.

All techniques will benefit from borehole data, and they are absolutely necessary for a quantitative estimation of relevant parameters.

INTRODUCTION

The prediction of shallow gas is an important task which will make drilling operations safer and can also contribute to a reduced rig time. The most important thing is to be able to predict this shallow gas with reliability so that blow-out can be avoided in any drilling operation.

The strategy for shallow gas prediction from seismic data will change from place to place depending on the water depth, the depth of the gas pocket below the sea floor, the velocity gradient from the sea floor to the gas pocket and the length of the seismic streamer.

The most significant seismic parameters which can be applied in the prediction of shallow gas are the seismic velocity, the seismic response and the absorption of seismic energy. In this paper we will demonstrate how these parameters can be analysed and used in connection with real data. The procedures are simple, and since they are based on the most fundamental parameters in seismics, they happen to work in practice.

THE SEISMIC VELOCITY

The seismic velocity is amongst other factors dependent on the pore fill. Figure 1 shows the seismic velocities as a function of porosity for a shallow sand which is water and gas saturated respectively. As seen from the figure, the P-wave velocity is most strongly affected by the pore fill. This is mainly due to the great difference between the incompressibility for water and gas. Figure 2 shows how the incompressibility or bulk modulus changes with gas saturation in the pore fluid. The difference in S-wave velocity for a water and gas saturated sand (see Figure 1) is due to the difference in bulk density.

It may be difficult, or impossible, to estimate the seismic velocity from reflection data with sufficient accuracy to be used as an indicator for shallow gas. Shallow gas is frequently associated with high amplitude anomalies in the seismic section. The problem in reflection seismology is to find a relevant reflector below this anomaly with sufficient response to obtain a good estimate of the interval velocity in the anomalous zone.

For shallow depths, refracted arrivals are superior to reflected arrivals in detailed velocity analysis. Berge et al.(1983,1985) have demonstrated the practical use of refracted arrivals in marine seismic data. In this paper we will demonstrate some additional procedures which are simple, but in many cases effective with respect to shallow gas studies.

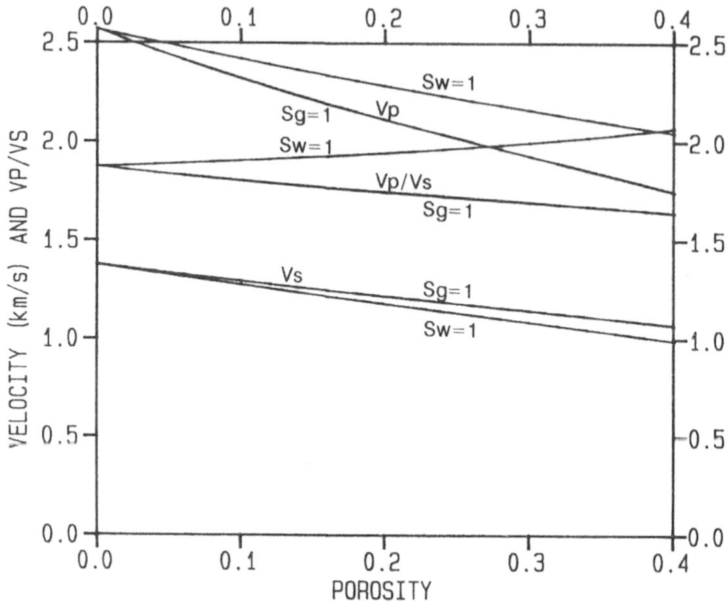

Fig 1. Seismic velocities as a function of porosity for a shallow sand.

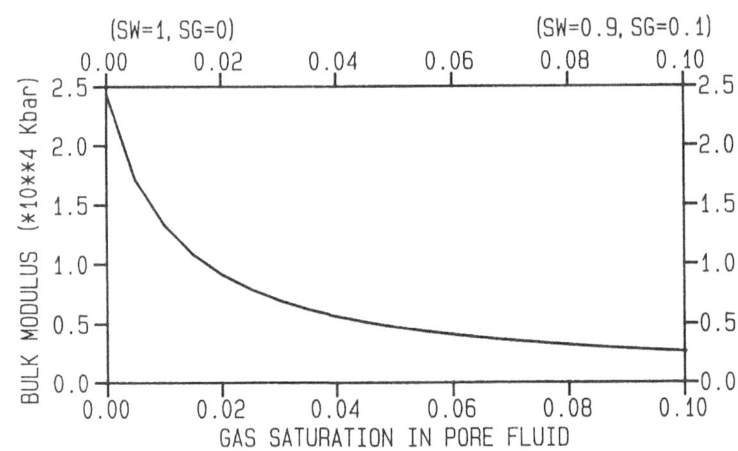

Fig. 2. The incompressibility of the pore fluid as a function of gas saturation

Figure 3 shows a CMP-gather, and in Figure 4 the same data are presented with reduced travel-time. The reduced travel-time $t_R(x)$, is defined by

$$t_R(x) = t_{obs}(x) - x/V_R \quad (1)$$

where $t_{obs}(x)$ is the observed travel time, x is the offset and V_R is the reduction velocity which can be selected. If a seismic record is plotted according to equation (1), then refracted arrivals with apparent velocity equal to V_R will appear flat. Thus relief on the refractor or changes in the refractor velocity produce apparent relief on the event.

Figure 5 shows first arrivals (refracted waves) where a reduction velocity of 3080 m/s is applied. The event appears flat, which means that the refractor velocity is close to 3080 m/s. The offset interval shown in Figure 5 is 2111 - 3111 m, and the data are taken outside an assumed shallow gas pocket. Figure 6 shows corresponding data which represent the shallow gas. As seen from the figure, there is a slope of the first arrivals indicating that the velocity is less than 3080 m/s. The data in Figure 6 have also been reduced by a velocity of 2850 m/s, resulting in a flat event as shown in Figure 7. This indicates that the gas has caused a reduction of the velocity of about 230 m/s.

The change in refractor velocity will also change the total travel-time of the refracted arrivals. This is shown in Figure 8 where the seismic traces with a constant offset of 2136m and

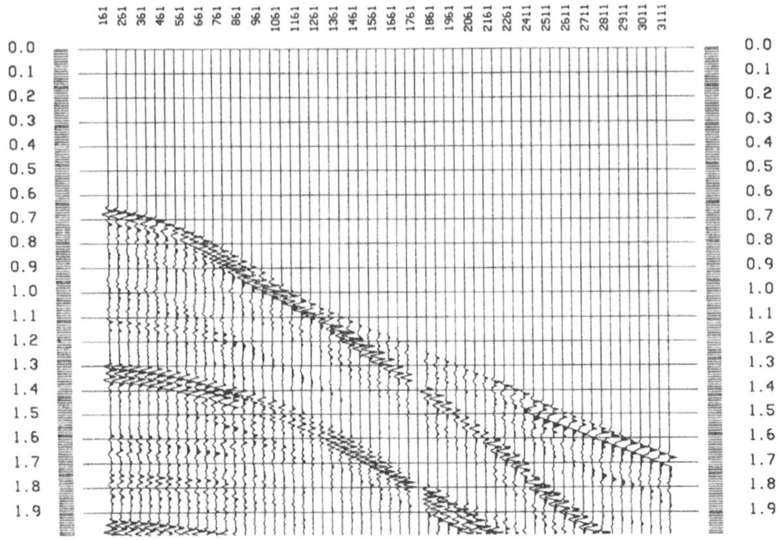

Fig.3. An ordinary CMP-gather.

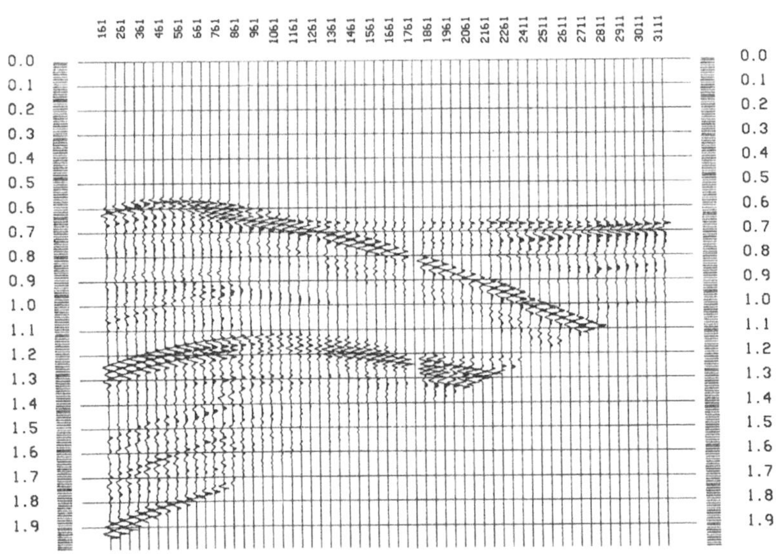

Fig.4. The CMP-gather in Fig.3 with reduced traveltime (V_R=3080m/s).

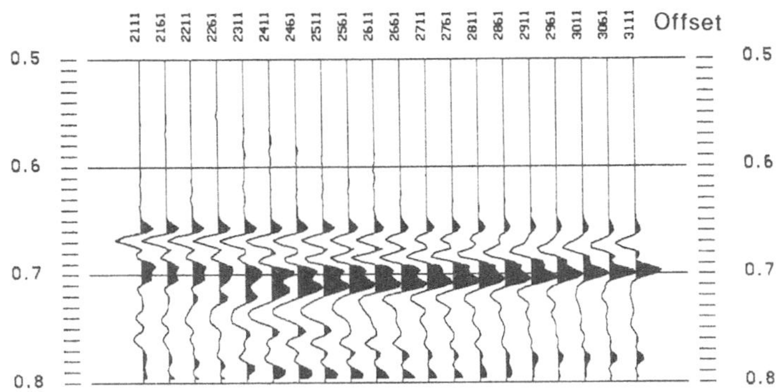

Fig.5. First arrivals with reduced traveltime (V_R= 3080m/s). Outside the gas zone.

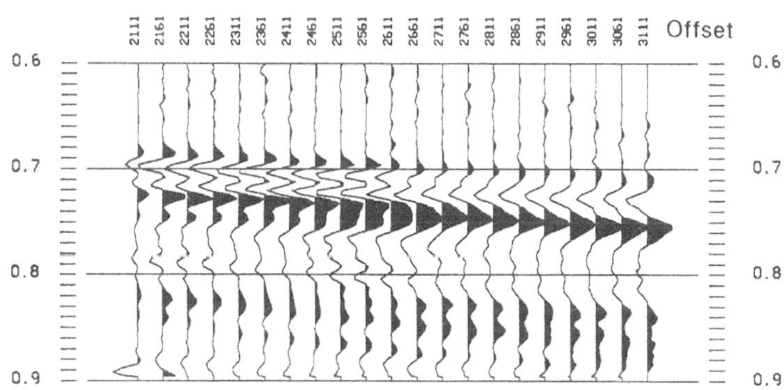

Fig.6. First arrivals with reduced traveltime (V_R=3080m/s). Within the gas zone.

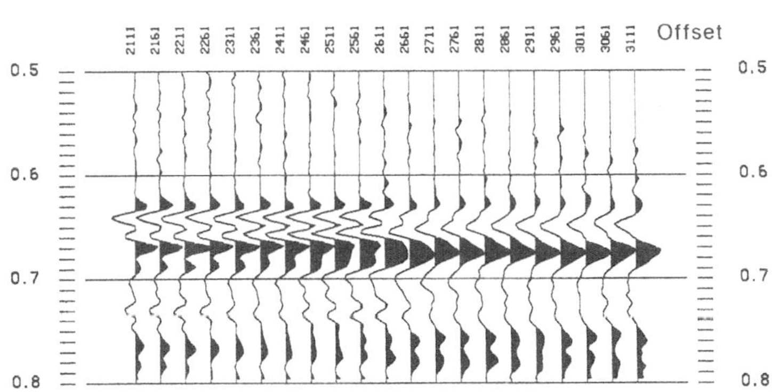

Fig.7. Same data as in Fig.6, but with V_R=2850m/s.

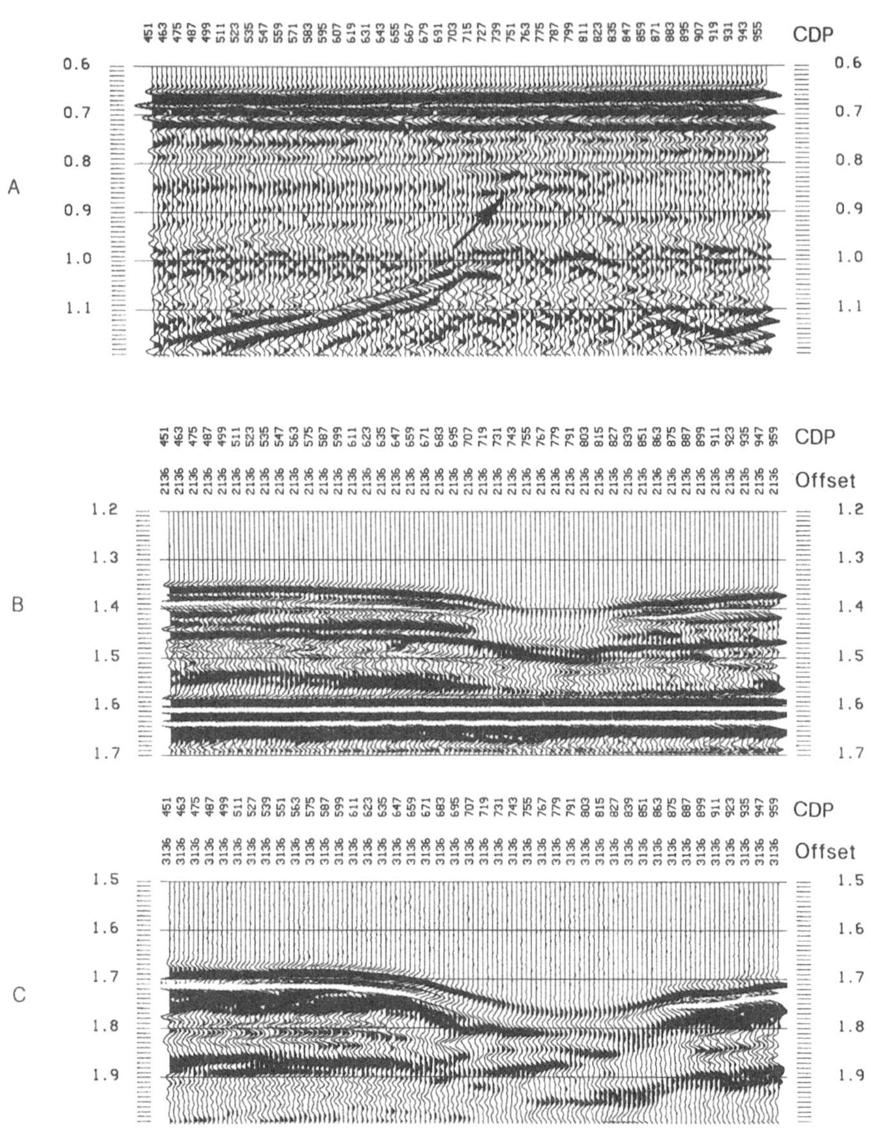

Fig.8 A) Near trace plot showing a shallow seismic anomaly.

B) Constant offset plot. True amplitude.

C) Constant offset plot. True amplitude.

3136m are plotted across the gas anomaly. As seen from the figure, the change in total travel-time is significant, and this method is convenient for mapping the lateral extent of the anomalous zone in terms of velocity variation.

ABSORPTION OF SEISMIC ENERGY

Another parameter which is indicative of gas is the absorption of seismic energy. In the case of partial saturation, the presence of the gaseous phase allows fluid within a pore or crack to flow freely under pure compression. Attenuation for compressional waves is much greater under this condition than in the fully saturated case. Laboratory investigations confirm this behaviour (Winkler and Nur, 1979; Spencer, 1979; Frisillo and Stewart, 1980).

In the case with a small amount of water in the pores where we have the shallow gas, we should expect to observe a drastic change in absorption of the seismic energy when we move into the gas zone. An example of this is shown in Figure 8. The attenuation of the refracted arrivals is shown for constant offsets (2136m and 3136m) where the seismic traces are displayed in true amplitude.

Successful use of seismic velocity and attenuation for prediction of shallow gas requires that refraction data from the relevant depths are recorded. This implies that in most cases a long streamer must be used during the data acquisition.

SEISMIC RESPONSE

The seismic response from the top of a sand layer will change with porosity and pore fill, and the angle dependent reflectivity is controlled by the V_p/V_s-ratio for a shallow sand with gas and water as pore fill, respectively. The physical model of the sand was established from borehole data close to the seismic anomaly in Figure 9 which was assumed to be caused by shallow gas. The physical parameters defining the sand are given in Table 1.

TABLE 1. Physical parameters of the sand.

Inkompressibility (k)	10.8GPa
Shear modulus (μ)	5.0GPa
Matrix density	2.65 g/cm^3
Formation pressure (assumed)	20 MPa

Pore aspect ratio spectrum:

Aspect ratio	Concentration
1.0	0.85
0.1	0.15

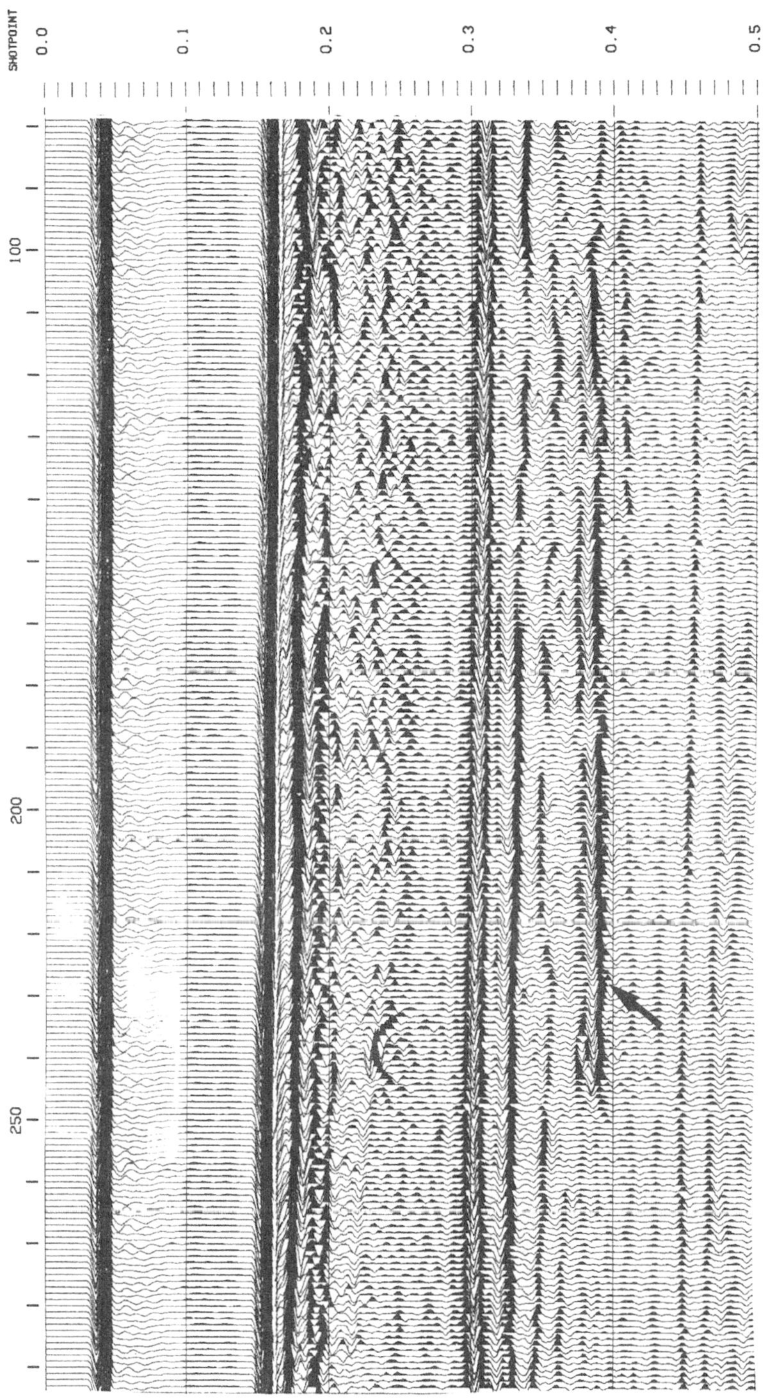

Fig. 9. Near trace plot showing a shallow seismic anomaly.

TABLE 2. Seismic model.

Depth (m)	V_p (m/s)	V_s (m/s)	Density (g/cm^3)	
0				
	1480	0	1.03	
132				
	1848	695	2.03	
290				
	2007	813	2.07	
332				
Sand layer	2137	1067	2.13	$S_g=0, S_w=1$
with seismic	1888	1125	1.85	$S_g=1, S_w=0$
anomaly	1874	1058	2.12	$S_g=0.05$, $S_w=0.95$
338				
	2103	1046	2.10	
350				
	2180	930	2.12	
356				
	2103	920	2.10	

The seismic model is defined in Table 2. The sand layer associated with the seismic anomaly is presented with parameters representing three different pore fill conditions. The seismic responses associated with these three models are shown in Figure 10. As seen from the figure there is a great difference between

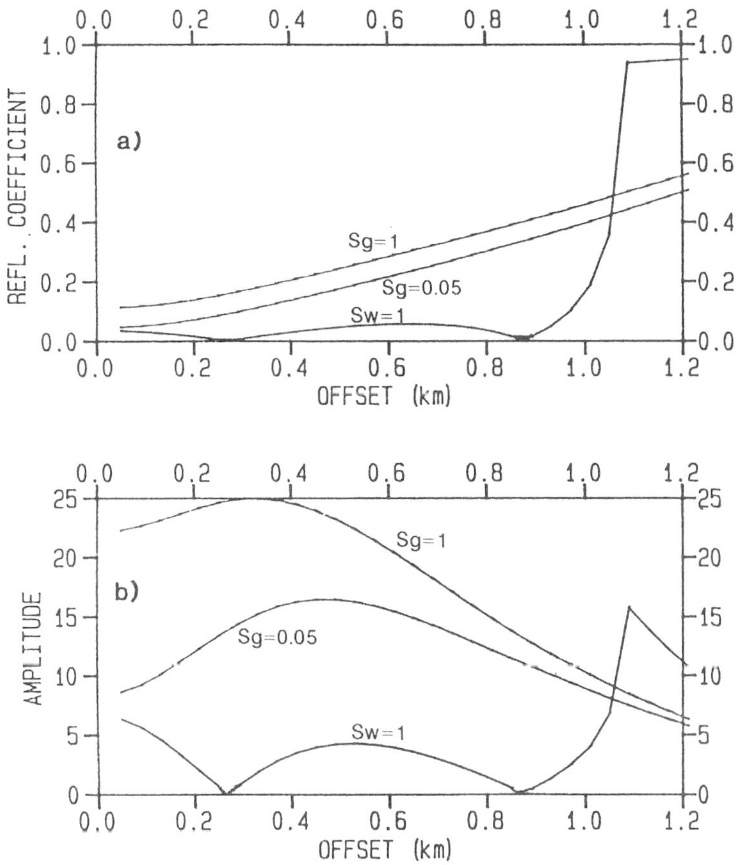

Fig.10 a) Reflection coefficient and

b) apparent seismic response from the top of the sand.

the reflection coefficient as a function of offset and the apparent seismic response (amplitude variation along the surface). In the latter case effects such as spherical divergence, absorption and transmission loss are included.

Figure 10 shows that there is a significant difference between the seismic response from the water saturated sand and the gas saturated sand. Even a small amount of gas (5%) results in a completely different seismic response compared to the water saturated sand. For the water saturated sand, polarity change occurs along the streamer at an offset of about 250 m and at an offset of about 850 m. These changes of polarity will have a destructive effect on the stacked signal, and a weak reflector may be wiped out during the stacking process.

The seismic responses in Figure 10 are displayed in terms of synthetic records in Figure 11 with the stack response to the right of the figure. As seen from the figure, the stack response from the water saturated sand is weak compared to the stack response from the gas saturated sand. From the near trace section in Figure 9 there is no indication of polarity change when we move away from the anomaly - as would be expected if $S_w=1$ (see Figure 10). However, a small amount of gas (of the order of 1 %) may be present, resulting in a very weak response and no change in polarity. The apparent seismic response for different gas saturations is shown in Figure 12, and for low gas saturations we see that the seismic response becomes close to zero for $S_g=0.01$ for small offsets.

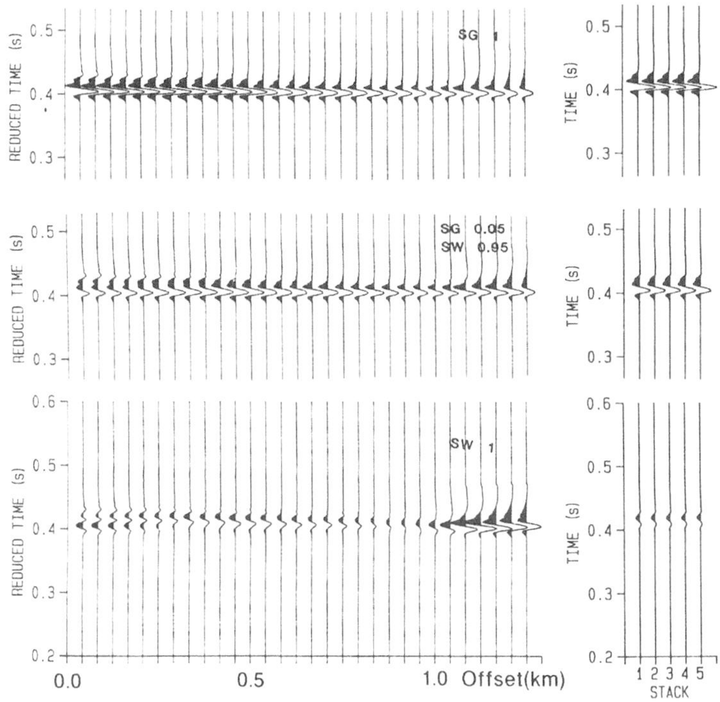

Fig.11 Synthetic data including reflections from the boundaries below the top of the sand. The stack response is shown to the right.

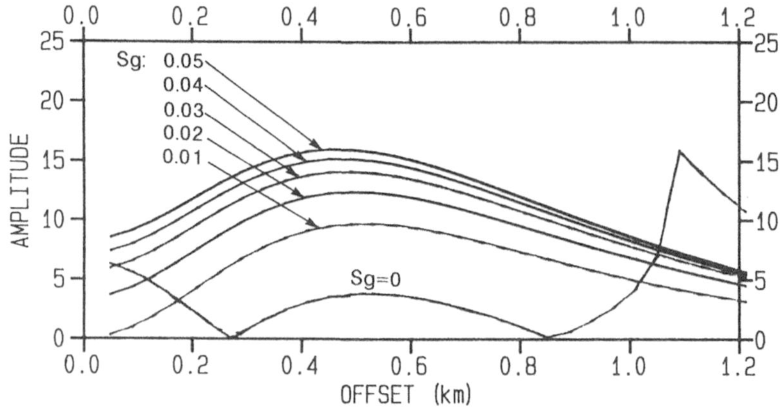

Fig. 12 Apparent seismic response from the top of the sand for different gas saturations.

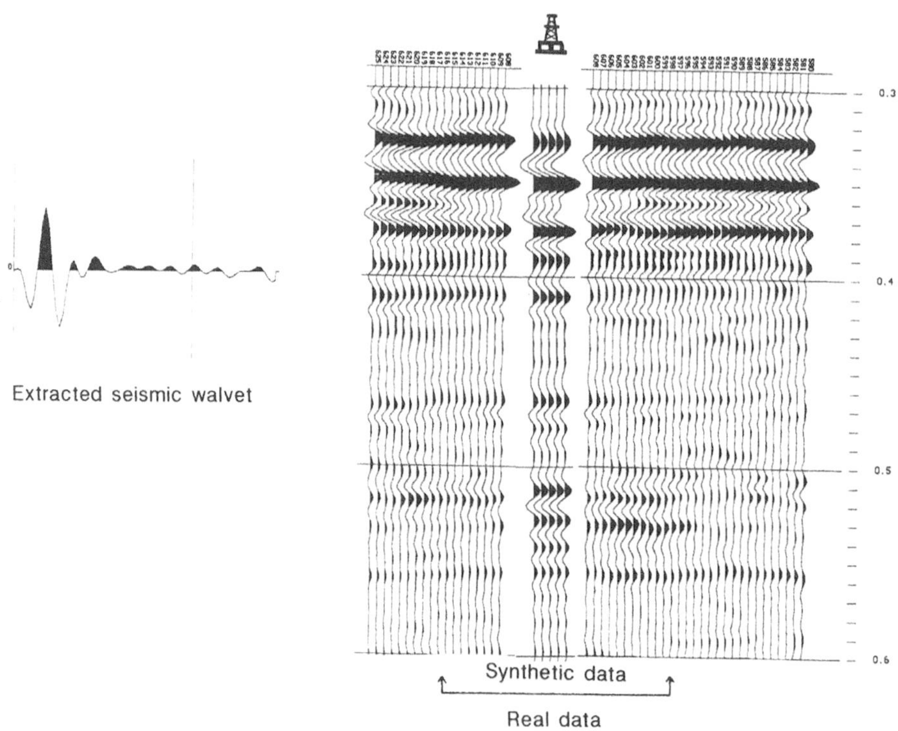

Fig. 13 Synthetic data and real data at the well position.

INVERSION

Borehole data available in the area can be used together with the seismic traces at the well position to extract the seismic wavelet. In connection with the seismic anomaly in Figure 9, borehole data were available from a well outside the anomaly. Figure 13 shows the extracted seismic wavelet and the synthetic seismic trace from the convolutional system model.The synthetic trace is displayed at the well position for comparison with the seismic section through the well. The agreement between the synthetic data and the real data is good.

The seismic wavelet is used in an inversion procedure across the seismic anomaly. The inversion removes effects of wavelet interferences and contributes to an increased resolution. The result of the inversion is a reflectivity time series as shown in Figure 14. After inversion the seismic anomaly appears as two strong reflections with opposite signs as shown in Figure 14.

The data in Figure 14 were converted to acoustic impedance. The variation in acoustic impedance can be interpreted in terms of variation in porosity and/or variation in porefill. The acoustic impedance as a function of porosity and porefill can be calculated from the parameters in Table 1, and an example is shown in Figure 15. The low acoustic impedance in connection with the anomalous zone is a strong indicator of gas as pore-fill when compared to the data in Table 2 and Figure 15.

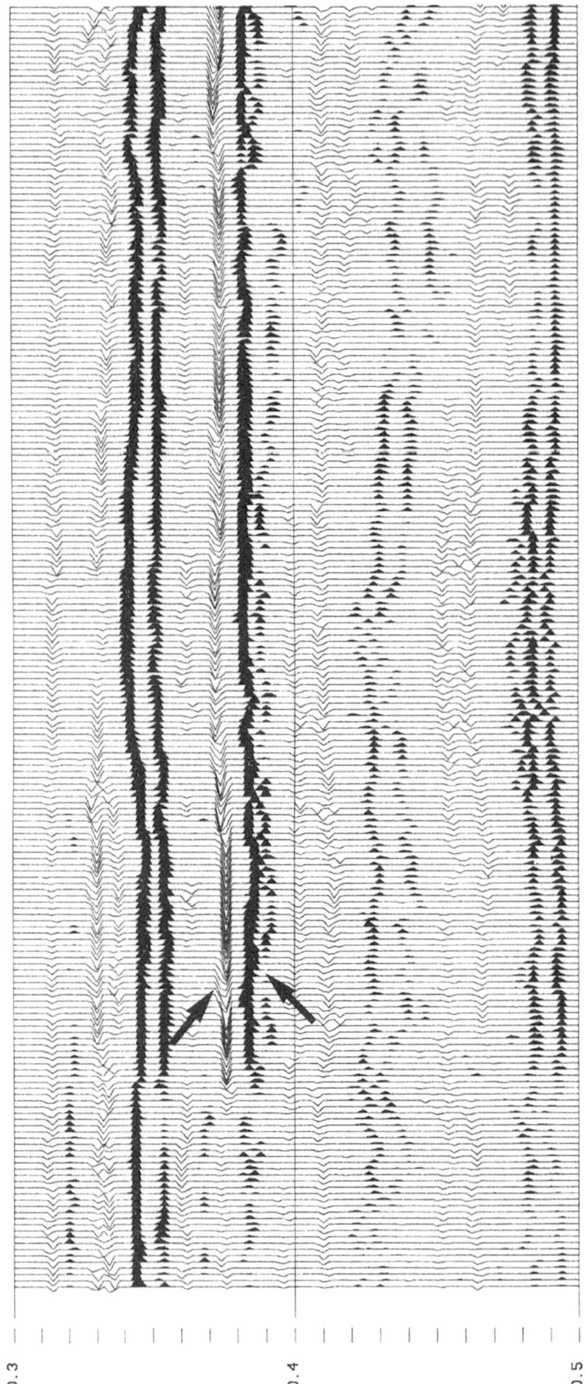

Fig. 14 The result of the seismic inversion.

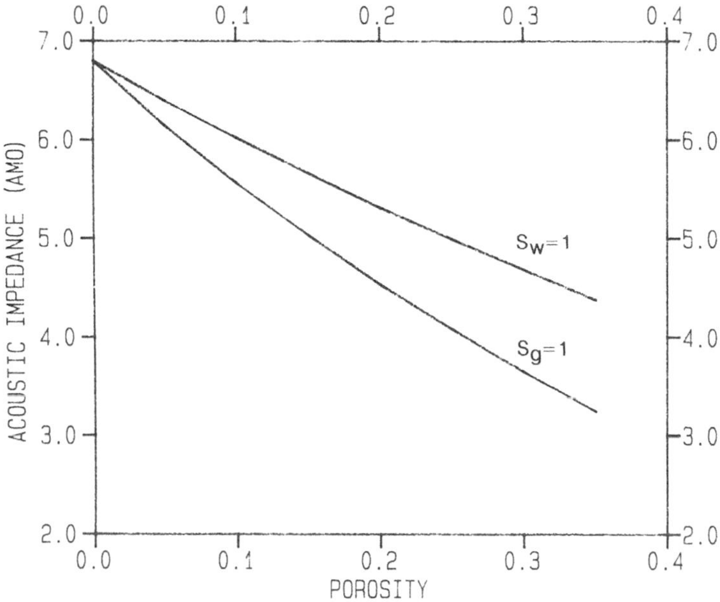

Fig. 15 Acoustic impedance calculated on the basis of the data in Table 1.

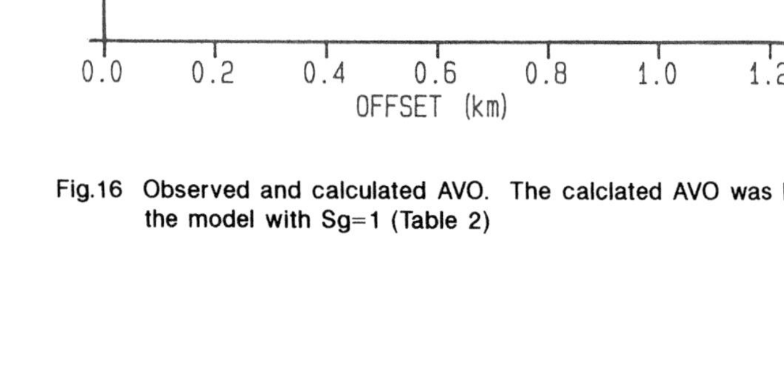

Fig.16 Observed and calculated AVO. The calclated AVO was based on the model with Sg=1 (Table 2)

AMPLITUDE VERSUS OFFSET

Amplitude versus offset (AVO) has a great potential in seismic exploration and lithological prediction. The methods are still under development, and a more effective use of this technique belongs to the future. However, from Figures 10 and 11 we can imagine that this technique can give valuable contributions in shallow gas prediction.

Amplitude versus offset have been estimated from the field data over the seismic anomaly. The data suffer from interference effects and do not allow detailed interpretation. Figure 16 shows an example with observed amplitude versus offset and calculated apparent seismic response based on the model with $S_g = 1$ (see Table 2). As seen from Figure 16, AVO also supports gas as pore fill.

CONCLUSION

Different techniques for prediction of shallow gas from seismic data have been presented in this paper. Since prediction of shallow gas is essentially a prediction of pore fill, we have chosen to focus on parameters which are significantly affected by a change in pore fill. If all the techniques could be applied at the same site, the result should be a reliable prediction.

The attenuation of seismic energy is a promising parameter, but it requires that refracted waves are recorded. The AVO technique is believed to be a powerful prediction technique for the future, but at the present time more development is needed. This technique also has the potential of estimating the saturation under favourable conditions.

Single trace inversion is also a promising technique, and with further development and linked to velocity modelling it will become a powerful tool in shallow gas prediction.

Direct use of seismic velocities in shallow gas prediction requires refracted arrivals. These data are of special importance where no borehole data are available.

All techniques will benefit from borehole data, and they are absolutely necessary for a quantitative estimation of relevant parameters.

REFERENCES

Berge, A.M., and Beskow, B.(1985). A method to determine the velocities in the seafloor and near-surface sediments, Geophysical Prospecting 33, 377-399.

Berge, A.M., Kanestrøm, R. and Pedersen, L.M. (1983). Detailed seismic velocities in sediments along a 250 km profile in the Barents Sea, First Break 1, 8, 9-17.

Frisillo, A.L. and Stewart, T.J.,(1980). Effect of partial gas/ brine saturations on ultrasonic absorption in sandstone, J.Geophys.Res. 85, 5209-5211.

Spencer, J.W.,Jr.(1979). Bulk and shear attenuation in Berea sandstone: The effects of pore fluids, J.Geophys. Res. 84, 7521 - 7523.

Winkler, K., and Nur, A., (1979). Pore fluids and seismic attenuation in rocks, Geophys. Res. Litt. 6, 1-4.

Part V
Procedures, Regulations and Guidelines

QUANTITATIVE GAS HAZARD ANALYSIS: PRESENT LIMITATIONS AND THE WAY FORWARD

by P. Conn* and J. Arthur**

* Well Studies Limited, The Gatehouse, 2 Holly Road, Twickenham, Middlesex, TW1 4EE.

** Top-Hole Site Studies Limited, The Gatehouse, 2 Holly Road, Twickenham, Middlesex, TW1 4EE.

ABSTRACT

The analysis of high resolution seismic data for shallow gas has made significant advances in recent years but there still remain major limitations in the ability to positively identify gas, let alone be able to accurately estimate concentration or pressure. A number of features which are commonly related to the existance of gas have been identified and when two or more such features are present this is commonly taken as being a positive gas indication and a probability of gas presence is often estimated. Such probability estimates are largely arbitrary (often based primarily on lateral amplitude variations) and the accuracy of such estimates is dependent on the skill of the analyst and knowledge of local geology. This paper discusses a series of techniques which should lead to more reliable estimates.

Volume 25: Safety in Offshore Drilling, 235–256.

There are several ways in which gas identification could be improved given our current knowledge: more careful processing of site survey data including better amplitude control in the shallow section, detailed velocity analyses in the vicinity of well locations and migration of key lines; use of extra analysis techniques such as residual wavefield and forward modelling for Amplitude Versus Offset; use of numerical analysis rather than qualitative estimation.

Implementation of some or all of the above techniques enables a better estimate of the probability of gas presence to be made but there still remains uncertainty especially in areas of irregular subsurface relief. It is proposed that in order to 'calibrate' gas probability estimates more research is needed relating known occurrence of shallow gas to the presence of gas related attributes in seismic data.

INTRODUCTION

Significant advances have been made in recent years in the basic evaluation of marine sites for the presence of gas in the top-hole sections of wells. However, there are still limitations to the quantitative analysis of the volume of gas present and even to the certainty of gas charge. In this context 'quantitative gas hazard analysis' refers to the use of numerical methods to interpret the presence of gas, define its probability and measure its areal extent and thickness.

A number of seismic gas indicators have been identified. These include amplitude anomalies, flat spots, phase shifts and frequency absorption. A positive indication of gas is assumed when two such indicators are observed together and an estimate of gas probability is made. Such estimates are often largely arbitrary and are very dependent on the skill of the analyst and the knowledge of local geology.

In order to improve the identification more effort in computer processing of digital data and greater use of numerical measurements are required. In particular the following points should be considered: more careful processing of site surveys with special attention to amplitude control in the shallow section, detailed velocity analyses in the vicinity of the proposed well, migration of key lines, use of extra analytical processes such as residual wavefield and Amplitude Versus Offset, forward modelling in instances where the nature of a possible gas indicator is uncertain and the use of numerical techniques (e.g. calibrated colour amplitude) rather than a qualitative estimate.

Current techniques are steadily improving and advances are greatly aided by the wider availability of powerful personal computers which are capable of interactive analysis and colour display. Nevertheless, it is considered that there will always be some uncertainty in the analysis, especially in new exploration areas, and more research relating observed gas occurrence, from well logging, to seismic expression is desired.

The techniques which are currently in use should always give indications of gas hazard when gas is present provided that they are applied correctly and carefully. Such a statement assumes that each stage of the gas hazard survey, from acquisition to final analysis, is correctly monitored to ensure that results output from each stage are of adequate quality.

The industry has now reached a point where the major problem has become not 'can we identify a hazard' but 'are all the possible hazards genuine'.

LIMITATIONS IN GAS HAZARD ANALYSIS

Limitations Related to Data Processing

The current techniques for shallow gas analysis have several perceived limitations. Possibly the most common limitations encountered are related to data processing. This is not to imply that data processing is incorrectly performed but rather to suggest where the approach to site survey data processing and analysis can be improved with extra effort.

Traditionally seismic data are plotted as black and white traces in a variable area with wiggle format. To study the data for gas indicators, the data has to processed to be relative amplitude. Even where the processing may be considered perfect the use of black and white displays with overlapping traces limits the resolution of the analysis both laterally and vertically. This is illustrated in Figures 1 and 2; both show a clear gas anomaly but the magnitude of the amplitude change across the anomaly cannot be calculated from the overscaled black and white plot but can be immediately calculated from the calibrated amplitude colour display. It can be observed that the magnitude of the amplitude anomaly is four times greater than the background reflection level along the horizon. (The anomaly is bounded by the 44298 contour whilst the background level seen at the left of the plot is limited at 10222). This type of numerical analysis of amplitude data is made much simpler by the use of calibrated colour plots and is virtually impossible if only paper plots of the sections are available.

Sometimes the energy source or data recording levels can vary during the recording of a site survey. (It may be that weather conditions worsened or that the source or receiver array depth varied as the boat steered in different

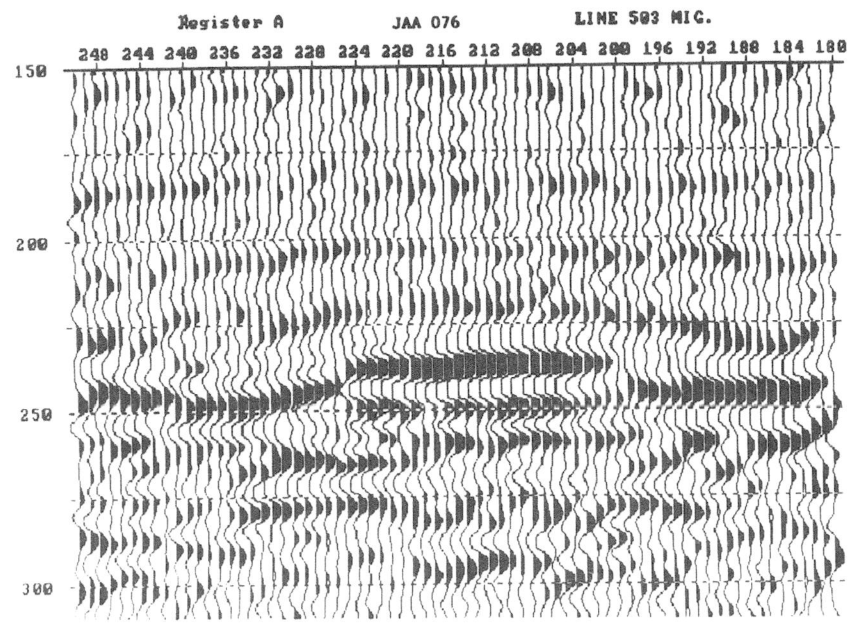

Figure 1. Black and White Variable Area/Wiggle display.

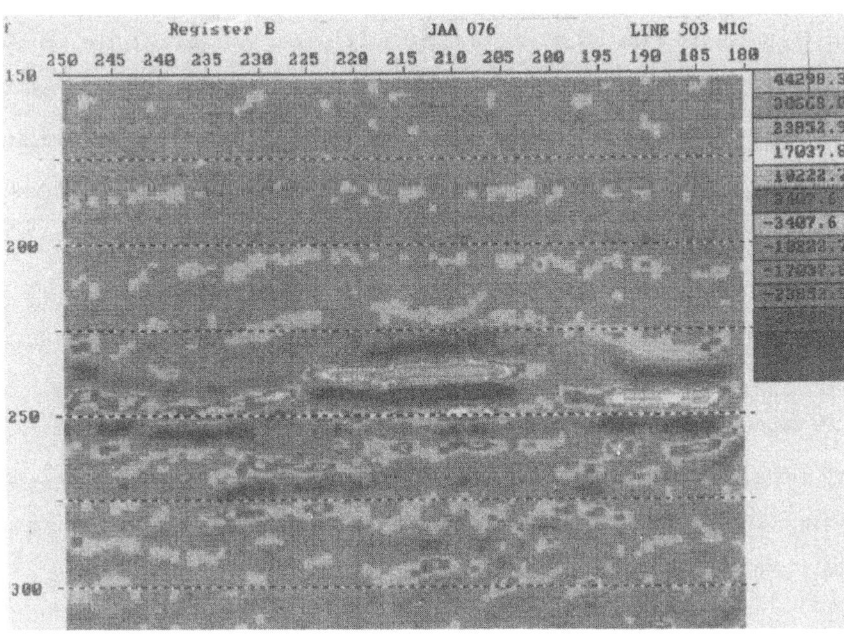

Figure 2. Calibrated Amplitude Colour Display (reproduced in grey scale).

directions.) When such a problem occurs it makes the use of numerical analysis techniques more difficult because it is not possible to assume the absolute amplitude of a given reflection from a fixed point will be constant on different lines even if all other factors are equal. Figure 3 is an example of amplitude mis-tie at an intersection.

It is common practice to record 48 trace data but to process it as an interleaved 24 fold section. If dead channels are present on the seismic cable the fold and average amplitude of the successive traces in the final interleaved section may differ. An example is shown in Figure 4. Such effects make quantitative interpretation of potential gas anomalies in the shallow section more difficult.

Another major problem that can affect the interpretation of the shallow section is inadequate gain control. It is sometimes the case that a gain recovery function is used that is derived from a single test panel that may not even be near the proposed well location. An example of inadequate gain control is shown in Figure 5 where a selected trace is shown with its original amplitude and after application of a revised amplitude recovery function. The relative amplitude of events along the time axis (i.e. in depth) is much clearer after correct amplitude recovery.

Limitations Related to Coverage

It is important to realise the value of site surveys. An inadequate survey can cost lives and prove very expensive. The planning of a suitable site survey should not be left to the last minute but should be ready in good time so that a full analysis of potential hazards can be made.

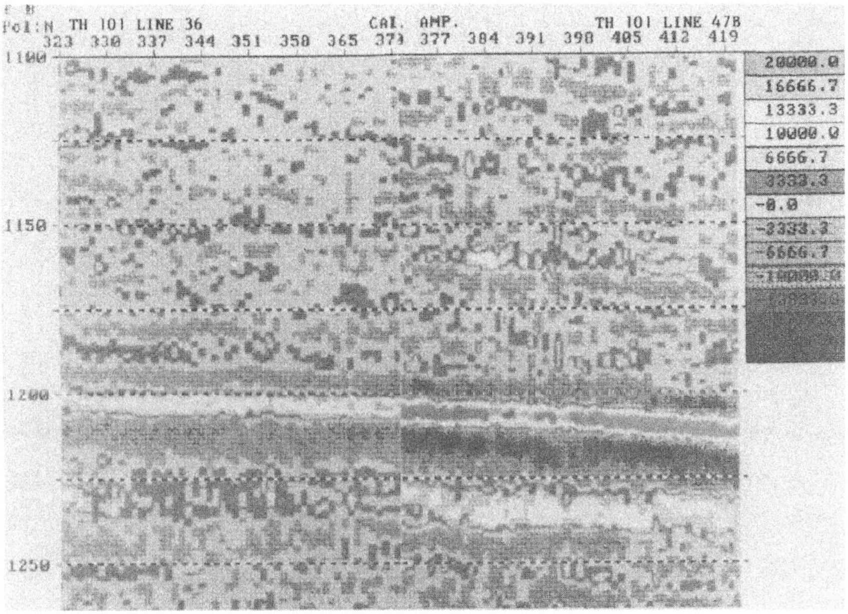

Figure 3. Amplitude Mis-tie at Line Intersection (Original in colour).

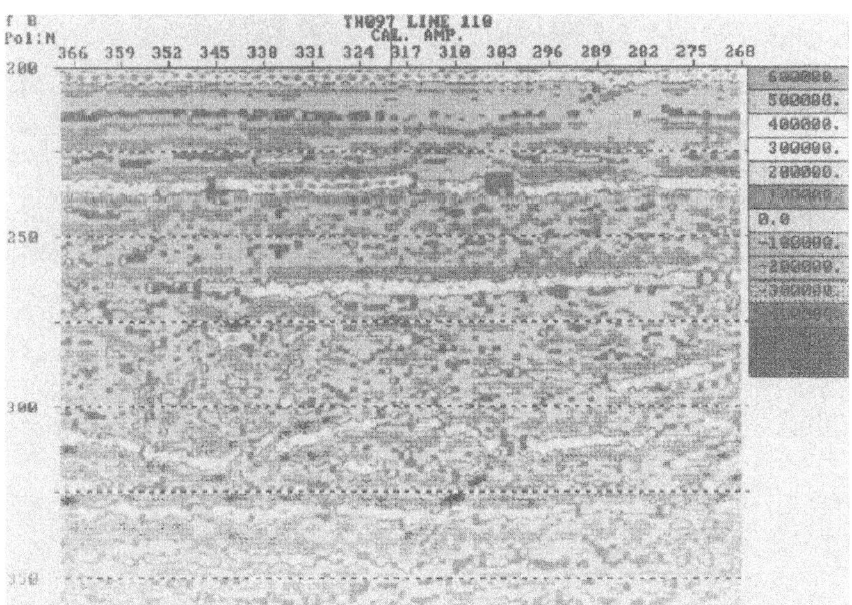

Figure 4. Successive Trace Amplitude Variation due to Dead Near Channel.

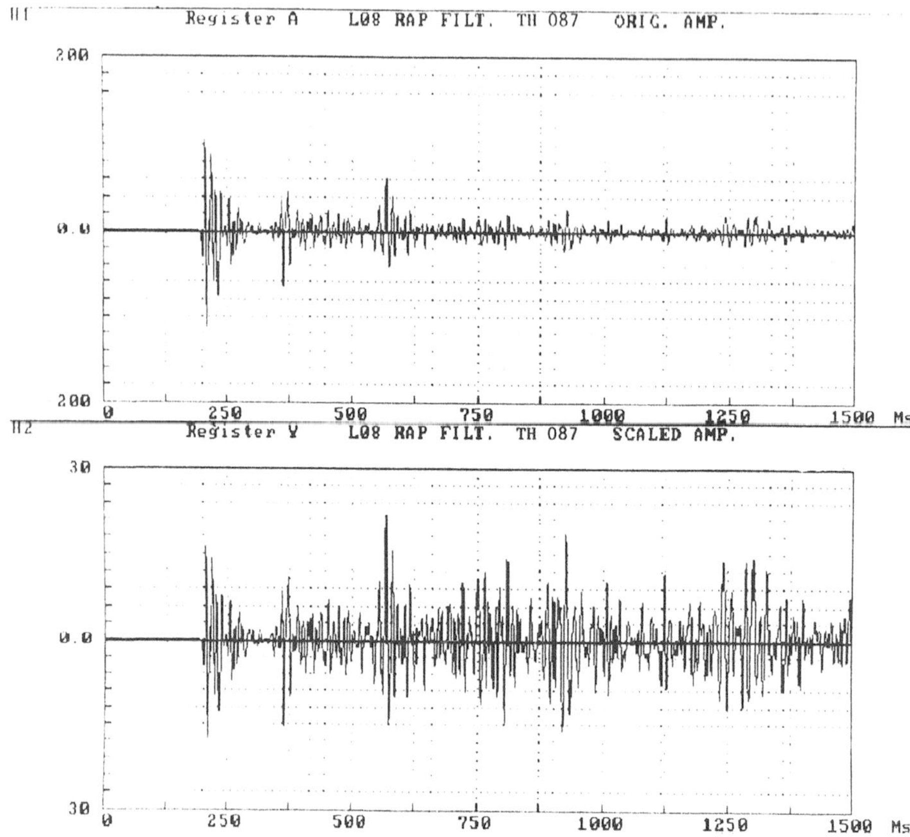

Figure 5. Example of Amplitude Recovery Correction

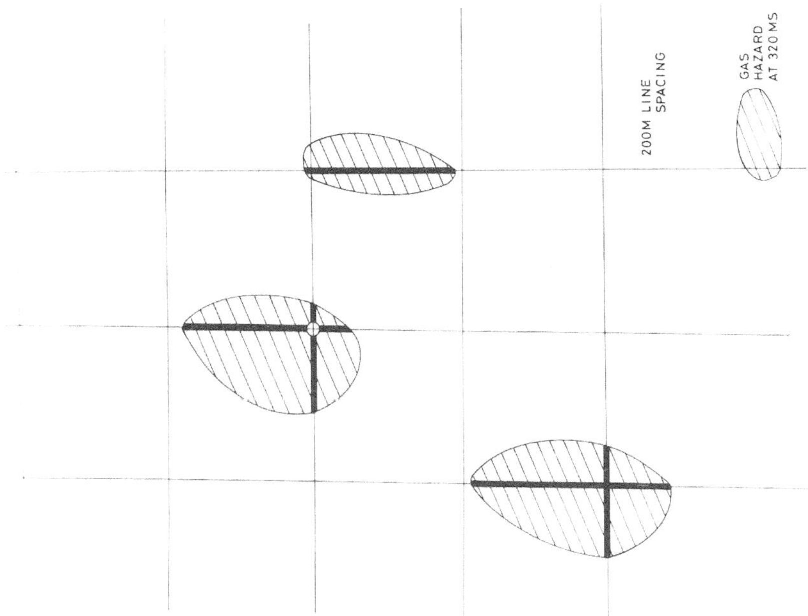

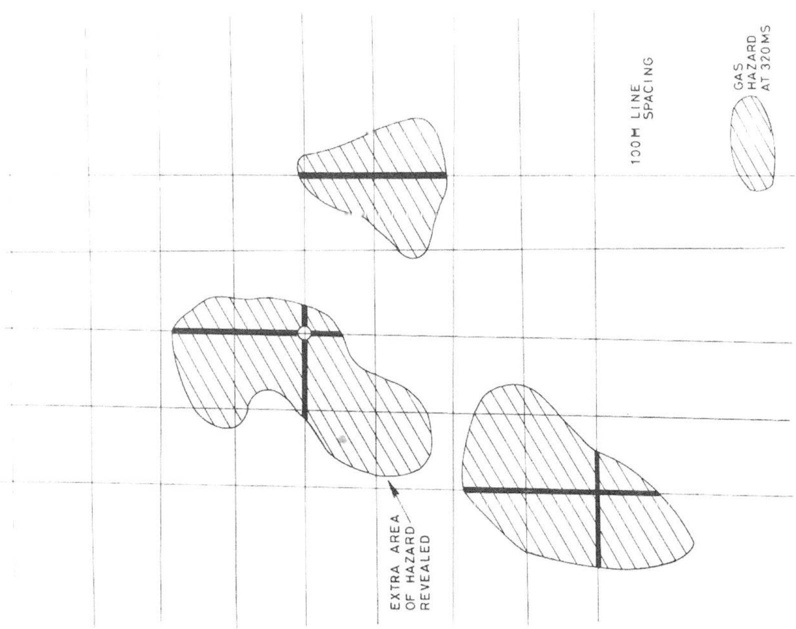

Figures 6 and 7. Interpretation of Gas Hazard with Different Line Spacing

The recent UKOOA guidelines are very useful for operators not experienced in setting up site surveys in UK waters but they cannot be regarded as totally definitive as each site is particular. It should be noted for example that a semi-submersible structure has different requirements for a site survey than does a fixed platform.

Past site surveys have often been inadequate to fully define the shallow structure and allow an understanding of the trapping mechanisms. This illustrated on figures 6 and 7 where the areal interpretation of a probable gas hazard at 320ms. has been significantly modified by the use of a 100m. grid rather than a 200m. grid. This example illustrates the improvement in interpretation possible with closer grid spacing. Figure 7 shows indications of a shoal-like structure with possible channels whereas Figure 6 appears to indicate completely isolated pockets.

Limitations Related to Field Geometry

Field geometry can have major effects on the resolution of the final data. Two aspects considered to be of particular importance are the effects of long offsets in CDP stacking and the frequency distortion caused by receiver ghosts.

The area of concern in the first case is the possible loss of spatial resolution and the modification of the apparent frequency absorption characteristics due to smearing where long offset traces are used in stacking. Figures 8 and 9 illustrate particularly the apparent change of absorption characteristics when the stack of near traces (1-12) is compared with the stack of all 48 traces on a 600m. cable. Wavefront healing has increased continuity of the event at 680ms. However the far traces that give the improved continuity have undershot the anomaly at 170 - 270ms. and have thus not suffered the same

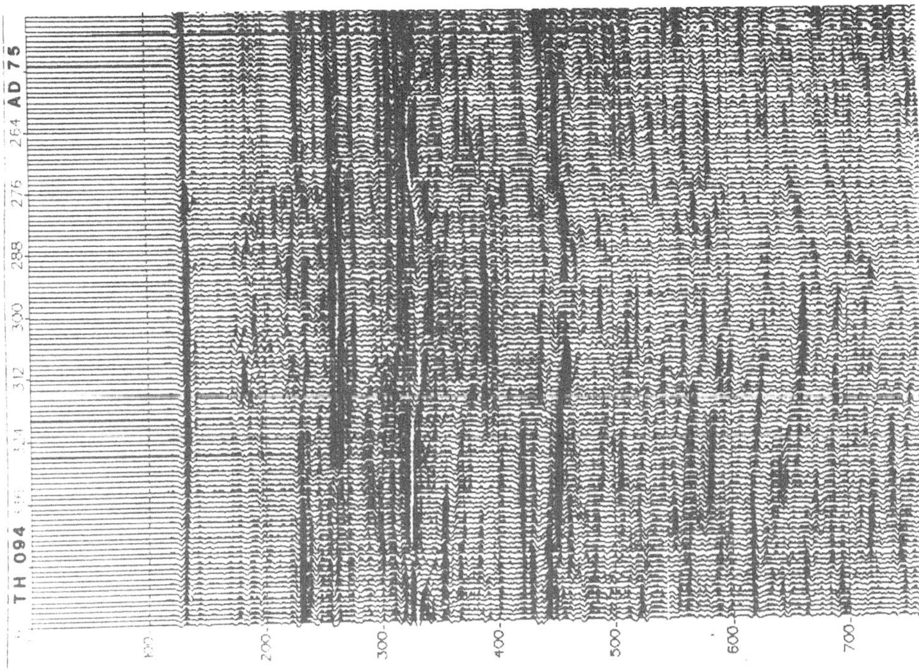

Figure 8. Stack of Traces 1 - 12 only.

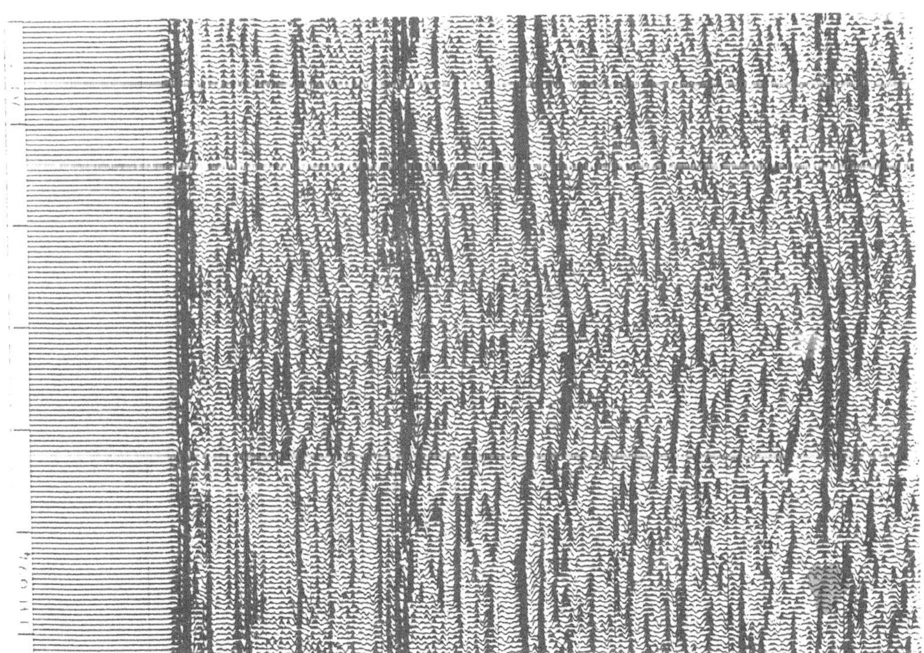

Figure 9. Stack of Traces 1 - 48 (timing lines 100 ms.).

absorption that the near traces have suffered. A study of the frequency spectra of both near and far traces in the affected CDP gathers reveals useful information on the extent of the anomaly but the final full fold section examined in isolation gives incorrect results when an instantaneous frequency display is used.

Vertical resolution may be lost because of the receiver ghost effect on a flat cable. The receiver ghost introduces a notch in the frequency spectrum of the recorded data and thus limits the vertical resolution. Recent tests by Seatex of Houston U.S.A., using a slant cable indicate that such an approach may give significantly improved results.

Geophysical Limitations

Many geological situations can give rise to anomalies that, at first sight, appear on seismic sections to be gas hazards. Pinch-outs, gravel beds, partly cemented sands and even limestone reefs under some circumstances demonstrate several apparent gas indicators. As lateral amplitude variation is usually treated as the most important gas indicator (the classic 'bright spot') it should be said that 'all that glows is not gas'.

Accurate volumetric analysis of gas content in a section is virtually impossible with p-wave methods. The behaviour of formation velocity and reflectivity with a gas-brine mixture for even low levels of gas (5-10 per cent) is very similar to that of a formation with a 100 per cent gas saturation (see Figure 10 from Domenico, 1976). The behaviour of s-wave velocity is theoretically proportional to the gas percentage but the changes are so small as to be insignificant under normal site survey circumstances.

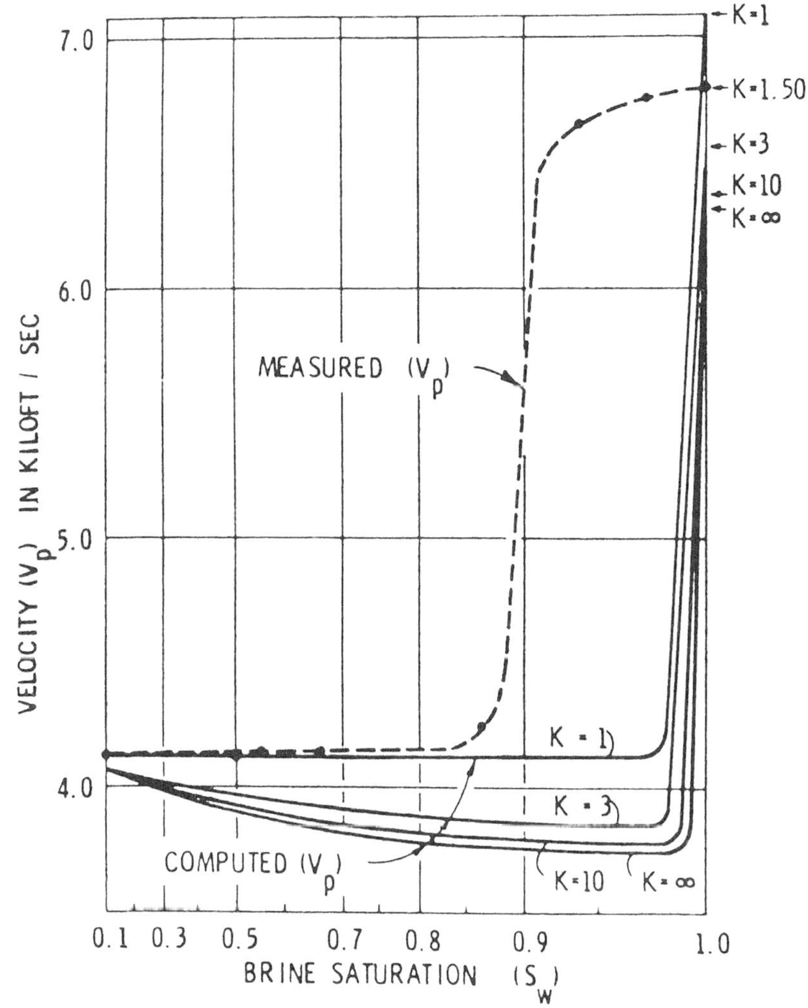

Figure 10. Measured and computed theoretical compressional wave velocities. (from Dominico (1976)

Over-pressure zones are very difficult to detect. Anomalously low velocities may be detected by detailed velocity analysis but it is difficult to distinguish between the effects of a low percentage gas-brine mixture and an over-pressure zone with current methods.

THE WAY FORWARD

Several aspects of current gas hazard evaluation can be improved to give more reliable data for numerical analysis.

Greater Care in Processing

As discussed earlier, particular care should be applied to handling dead traces. A knowledge of the time-offset amplitude decay across the receiver spread is needed to calculate correct scaling factors to compensate.

Analysis of amplitude decay should always be performed at several positions in order to derive an average function. One analysis point should be at the well location.

The use of a pre-stack spike removal process is desirable. Such a process could be based upon a median filter or a diversity stack.

Amplitudes of sections at line intersections should be matched. This can be done by the use of a series of constant scalars. The scalars can be computed by a least squares technique.

Velocity control can affect the section amplitude response as is shown by

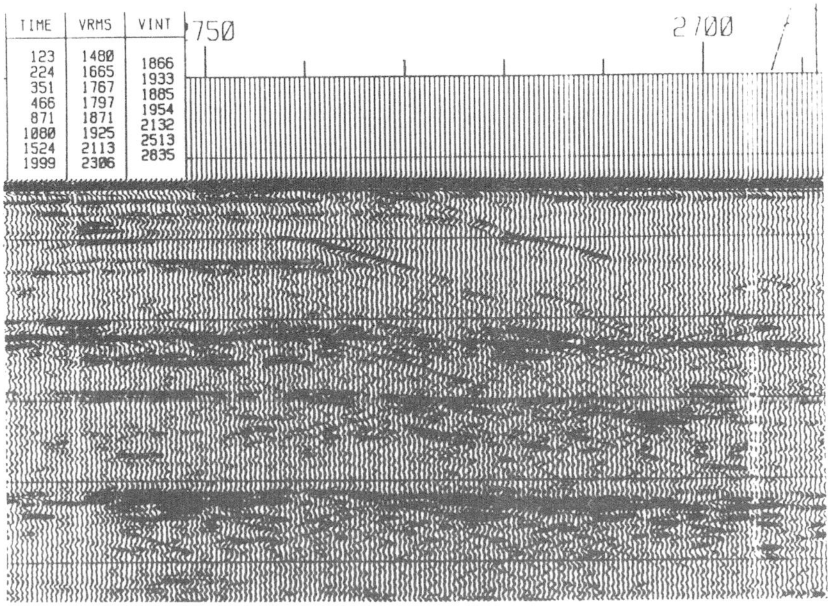

Figure 11. Intermediate Stack.

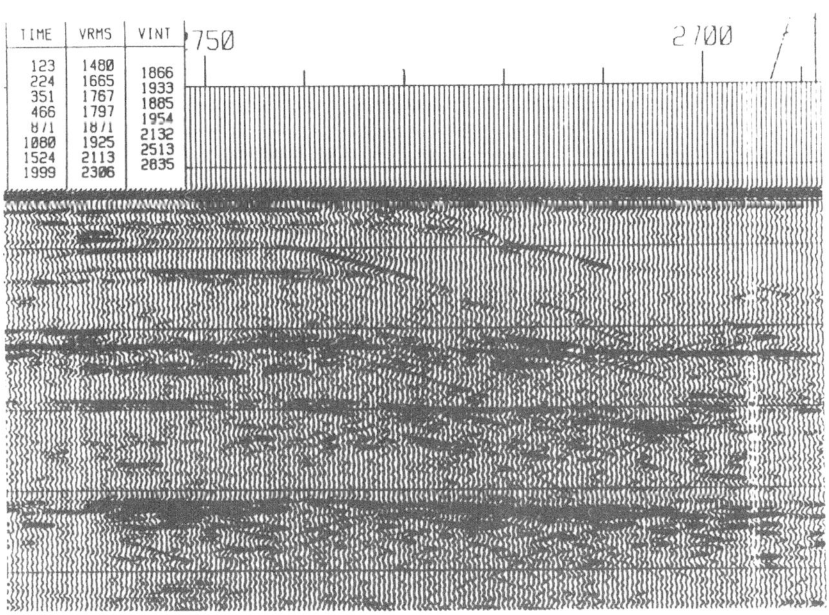

Figure 12. Final Stack with Revised Velocity Function.

Figures 11 and 12. The effects of a relatively small change in the moveout function applied to data from a 600m. receiver spread has caused key differences in the stack response from 320 - 700ms. The gas hazard analysis of the two sections, taken in isolation, would be different.

The strategy recommended for velocity analysis is as follows:

Perform a velocity analysis at the well location.
Use constant or variable velocity scans (CVS or VVS).
Plot the corrected gather at the centre of the analysis panel.
Use semblance plots to assist other methods not to replace them.
Increase the density of analyses.
Locate extra analyses at points of subsurface variation (e.g. sub-seabed channels)

Use of Additional Proven Analysis Techniques

Wider use of colour attribute plots such as Instantaneous Amplitude and Phase are especially useful (see Figures 13 and 14) but much information can be gained purely from Calibrated Amplitude colour plots.

Wider use of migration on key lines can not only reveal more structural detail (see Figure 15) but will also collapse diffraction hyperbola to give much better data for numerical analysis of Instantaneous Phase changes.

The use of the Residual Wavefield method can give valuable information on the origin of diffractions. The residual wavefield is defined as the energy remaining when coherent p-wave reflection energy has been removed from the data set. Figures 16 (Calibrated Amplitude) and 17 (Residual Wavefield) illustrate how a

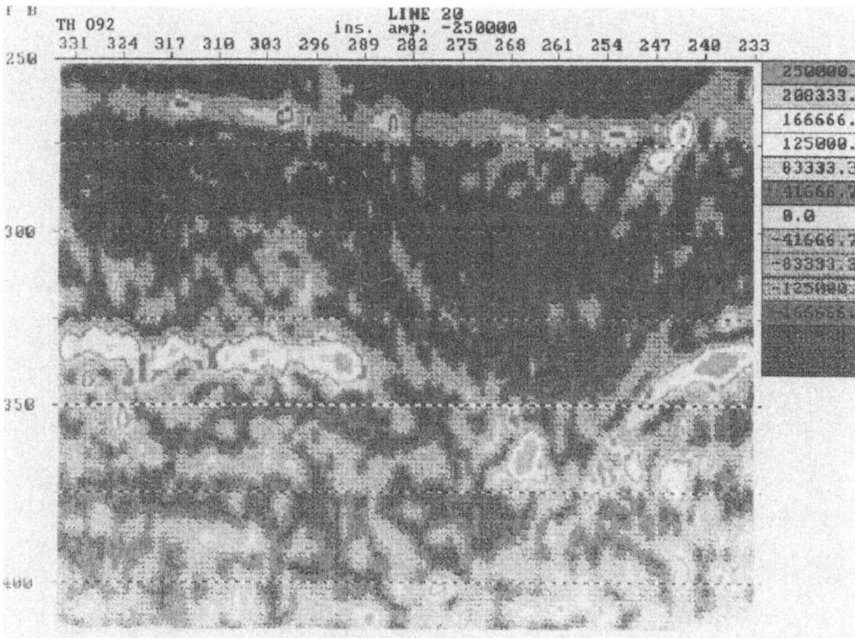

Figure 13. Instantaneous Amplitude Plot (Original in colour).

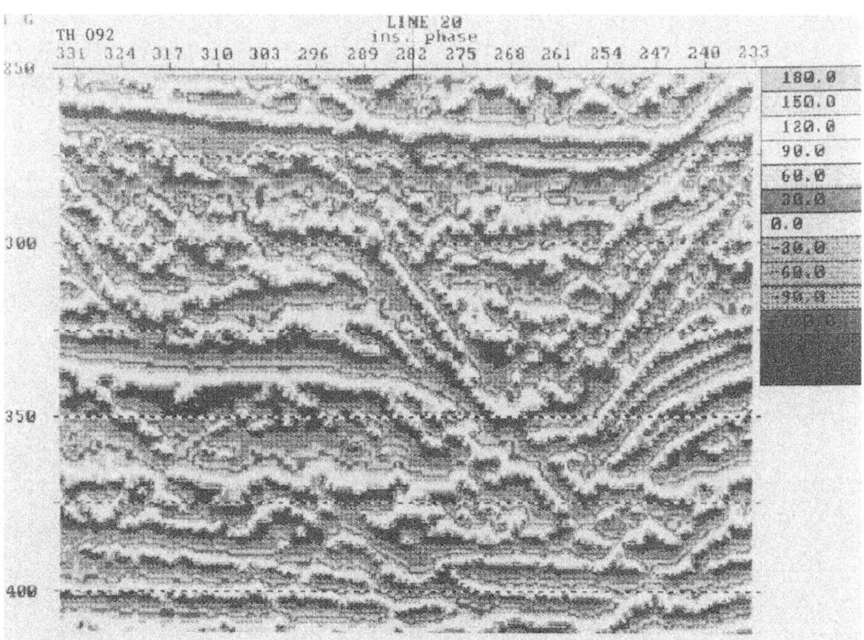

Figure 14. Instantaneous Phase Plot (Original in colour).

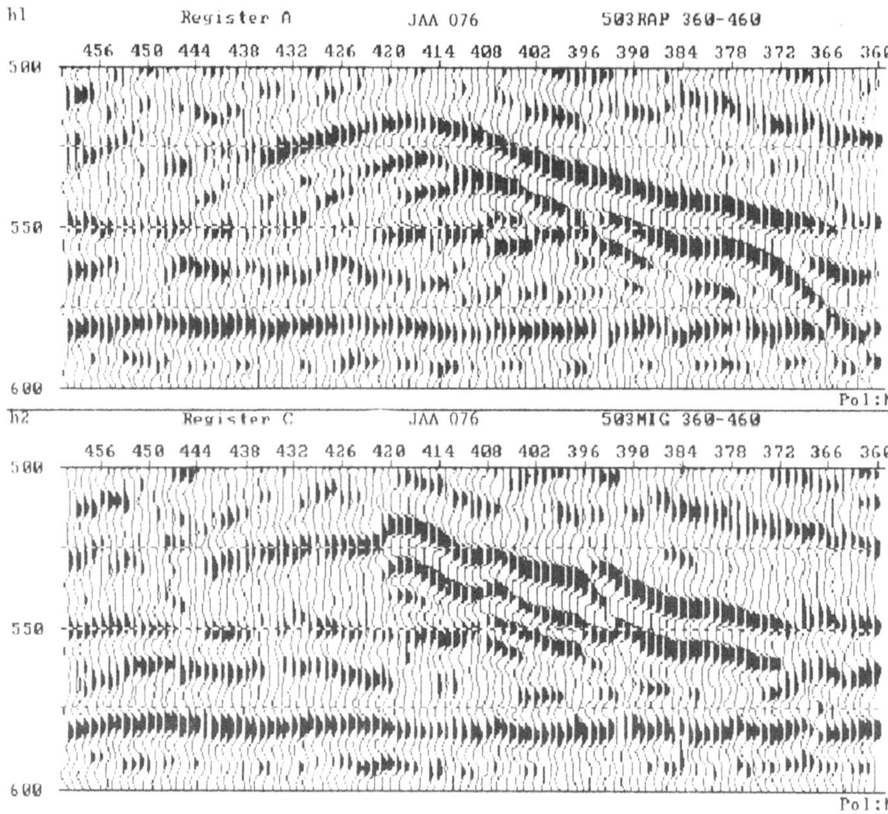

Figure 15. Mound Feature before and after Migration.

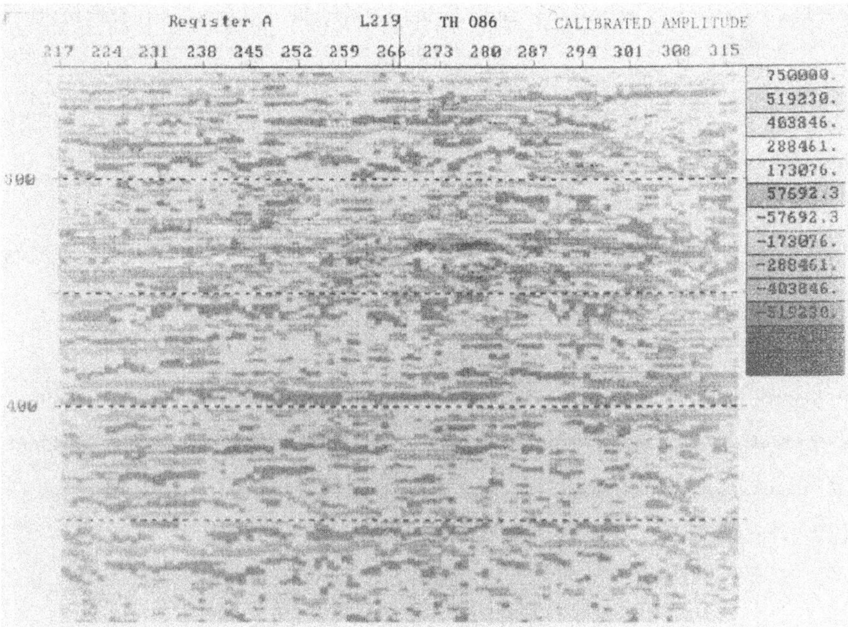

Figure 16. Calibrated Amplitude Plot (Original in colour).

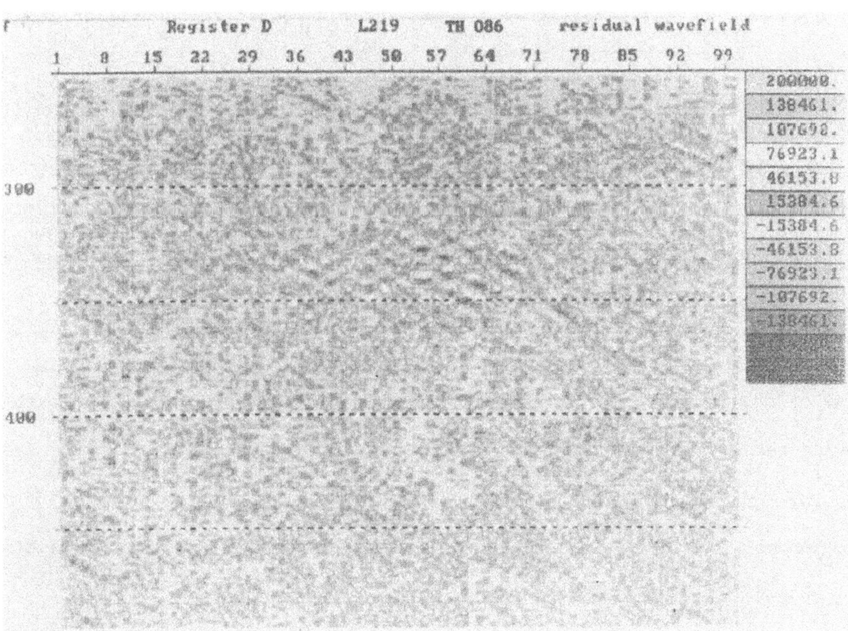

Figure 17. Residual Wavefield Plot from figure 16 data (Original in colour).

diffraction point can be clearly identified using the residual wavefield.

Site Specific Evaluation

It is vital to have an understanding of the lateral and vertical local geology so that the trapping mechanism can be understood .

When a probability of gas hazard is identified at a proposed well location it may be desirable to attempt to prove or disprove the hazard by a two-dimensional forward modelling study. Such an approach may save time and money that an unnecessary site location move would involve.

Further Evaluation of Techniques of Promise

AVO is seen as a technique of great promise for more positive identification of areas of gas charge.

Pulse shaping effects require more study. The use of calibrated amplitude plots can qualitatively give such information but numerical techniques including similarity functions could give a more precise evaluation. In key areas the pulse shaping study may need to be tied to forward modelling.

The use of a slant cable for acquisition can provide better resolved data thus improving the estimate of the depths of the gas layer boundaries.

3D site surveys have been demonstrated in Norwegian waters. The results presented appear to give clear indications of the volume of gas pockets.

Instantaneous frequency can give clear indications of the extent of gas. The degree of frequency loss may also aid volumetric gas analysis.

CONCLUSION

There still remains uncertainty in gas hazard analysis. The uncertainty can be reduced by careful processing with special care in amplitude and velocity control, and wider use of now proven techniques such as migration, colour attribute analysis and residual wavefield.

A site specific approach to site surveys can help define the area of gas charge but volumetric analysis of gas remains uncertain because of geophysical limitations. A well planned 3D site survey may prove to be the ultimate site specific technique for platform locations.

Quantitative analysis methods and the ability to define a gas probability on a sound numerical basis are considered vital.

Further research relating known gas occurrence to observed seismic indicators, including shaping effects and instantaneous frequency, is needed for better quantitative analysis.

References

Arthur, J.C.R., (1988). Offshore Drilling and Safety: The Role of Digital High Resolution Site Surveys. In Advances in Underwater Technology, Ocean Science and Offshore Engineering, Volume 16: Oceanology '88, Graham and Trotman, London, 197-201

Dominico, S.N., (1976). Effect of Brine-Gas Mixture on Velocity in an Unconsolidated Sand Reservoir. Geophysics, Vol 41, 882-894

Newman, P. (1990) Enhanced Seismic Methods for Marine Hazard Surveys. In Safety in Offshore Drilling-The Role of Shallow Gas Surveys, Graham and Trotman, London.

Moore, N., (1989) The Added Vertical Dimension of the Slant Cable; A Unique Approach to Greater Seismic Depth and Resolution. Private Communication. Seatex Service Inc., Houston, U.S.A.

Taner, M.T., Koehler F. and Sherrif R.C. (1979). Complex Seismic Trace Analysis. Geophysics, Vol.44 No.6, 1041-1063

UKOOA RECOMMENDED PROCEDURES FOR MOBILE DRILLING RIG SITE SURVEYS (GEOPHYSICAL AND HYDROGRAPHIC) - SHALLOW GAS ASPECTS

P.M. Walker

Shell U.K. Exploration and Production,
Shell-Mex House,
The Strand,
London WC2R 0DX

ABSTRACT

UKOOA recommended procedures for mobile drilling rig site surveys have been developed. The procedures are, at present, in a proof form but they have been reviewed by UKOOA members and other interested parties.

The procedures provide information for a standard approach to rig site surveys with emphasis on good oilfield practice.

To illustrate the approach taken by these UKOOA procedures, this paper highlights areas of the procedures of particular interest for shallow gas surveys. Standards to ensure acquisition, processing and interpretation of seismic data that is suitable for shallow gas detection throughout the UKCS are proposed. Data examples are provided to illustrate the sensitivity of seismic data quality to acquisition specifications and processing.

Volume 25: Safety in Offshore Drilling, 257–289.

One of the conclusions of the procedures is that the quality of the survey results cannot be taken for granted. Careful quality control by specialists is required throughout acquisition, processing and interpretation of the data.

Even with data acquired to the limits of available technology, shallow gas surveys should be viewed as a preventive measure, minimising the possibility of encountering shallow gas, and therefore reducing the risk to acceptable levels for rig owners and their insurers.

1. INTRODUCTION

1.1 Overview

Procedures have been developed by UKOOA for the geophysical and hydrographic aspects of mobile drilling rig site surveys in the UKCS. The procedures are, at present, in a proof form but they have been reviewed by UKOOA members and other interested parties. Relevant comments are currently being incorporated.

To illustrate the approach taken by these UKOOA procedures, the following text highlights areas of the procedures of particular interest for shallow gas surveys. Some examples are provided to illustrate the sensitivity of seismic data quality to acquisition specifications and processing.

As the UKOOA document has yet to be finalised, the following text may not represent faithfully the final content of the procedures.

1.2 Background to the Development of the Procedures

With the discovery of offshore oil and gas fields the site survey industry has grown considerably over the past two decades. The primary tools of site surveys are very resolute seismic survey techniques which have developed from relatively simple analogue single channel recordings to more complicated digital multi-channel recordings which require data processing ashore. As the usefulness of the survey data has increased, more reliance has been placed upon the results.

Rig site survey data are acquired for several objectives including shallow gas detection and prediction of soil lithologies, identification of seabed hazards, and measurement of water depth. Of prime importance is the use of shallow gas surveys to predict potential shallow drilling hazards. Sub-standard data and inadequate or improper use of survey results may have considerable financial implications in addition to jeopardizing safety. Despite these serious implications, no detailed procedures presently exist in the UKCS for performance of these surveys.

In November 1986, UKOOA members suggested to the UKOOA Surveying and Positioning Committee that procedures for the performance of rig site surveys would be desirable.

In November 1987 The UKOOA Seabed Survey Technical Sub Committee was therefore requested to develop procedures for the geophysical and hydrographic aspects of surveys of mobile drilling rig sites in the UKCS.

1.3 Objectives and Limitations of the Procedures

The procedures provide information for a standard approach to rig site surveys, with emphasis on good geophysical and survey practice. They cover both survey objectives and technical requirements. Limitations of equipment and techniques are also discussed briefly.

Even with data acquired to the limits of available technology and careful analysis of that data, the geophysical survey results are still an interpretation. In the case of shallow gas surveys, the results should be viewed as a preventive measure and not a definitive indicator of shallow gas. Lithology predictions are also not definitive and are improved by borehole or well data.

It is not intended that the procedures should be a text book on geophysics or geodetic positioning. There is already ample literature on these subjects and a reference list is provided for further reading.

1.4 Format and Content of the Procedures

The procedures have been divided into several brief sections which provide the basic information for the proposed minimum standards for rig site surveys. More detailed appendices provide background documentation in support of the standards and permit an understanding of the way in which results may be compromised if standards are relaxed due to operational or financial constraints.

Section 1 is an introduction to the procedures.

Section 2 lists the existing regulations and guidelines concerning the performance of rig site surveys.

Section 3 covers the important factors which should be considered in the planning stages of a rig site survey.

Sections 4, 5, and 6 cover shallow gas surveys, jack-up rig site surveys and semi-submersible rig site surveys. Survey objectives, recommended equipment, survey pattern and acquisition parameters, data processing, interpretation and reporting are discussed briefly.

The appendices provide background documentation for the procedures. They should be read in conjunction with the procedures for a full understanding of the implications of inadequate survey results.

2. EXISTING REGULATIONS AND GUIDELINES

2.1 General

Offshore site surveys should be undertaken within the confines of the following regulations and guidelines.

i) Petroleum Act Regulations, 1988.

ii) Minerals Workings (Offshore Installations) Act, 1971.

iii) The Offshore Installations (Construction and Survey) Regulations, 1974.

iv) The Health and Safety at Work Act, 1974.

v) Guidance on Foundations and Site Investigations for Offshore Structures. Department of Energy, January 1986.

vi) Offshore installations : Guidance on design, construction and certification. Section 14, Site investigations. Department of Energy, June 1988 (Draft Document).

vii) Seabed and Subseabed Data Required for Approvals of Self Elevating Platforms. Noble Denton International Ltd, 1987.

viii) Application for Consent to Drill an Exploration or Appraisal Well. Continental Shelf Operations Notice No. 11. Department of Energy, revised September 1989.

ix) Geophysical Surveys. Continental Shelf Operations Notice No. 33. Department of Energy, December 1976.

x) Notification of Seismic Surveys. Continental Shelf Operations Notice No. 35. Department of Energy, September 1977.

xi) Procedures Relating to the Notification of Vessels Intending to Anchor in the Vicinity of Pipelines or Other Subsea Installations. UKOOA Procedures Guide, agreed August 1986.

2.2 Regulations of Particular Interest for Shallow Gas Surveys

Of these regulations and guidelines, Continental Shelf Operations Notice No. 11 is of particular interest for shallow gas surveys. This states that applications for consent to drill should be submitted, by the operator for the licence, not less than 28 working days prior to expected start of operations date:

"The following information should be provided:-

(g) Results of the site survey which must address the possibility of shallow gas at the drilling site".

Exploration and appraisal wells are expected to be thoroughly surveyed at intermediate levels as well as over target horizon(s). The wireline logging programme must be adequate to properly evaluate any potential shallow gas horizons.

There is, therefore, a requirement for the operator to satisfy the Department of Energy that there is adequate shallow gas information available for each rig site.

3. SURVEY PLANNING

3.1 When are Surveys Required?

The objective of these surveys is to identify and map possible shallow gas accumulations.

Careful consideration must be given to the risks involved if surveys are not undertaken. In certain cases there may be good reason not to undertake surveys. For example, if the geology of an area is well known, there is no history of shallow gas accumulations and there are no potential gas sources or reservoirs, then a detailed review of such data may suffice. However, the Department of Energy, and the rig owners and insurers may not be satisfied with such regional information, so they must be consulted.

To decide on the need for a survey therefore, the available geological and geotechnical literature should be studied to review the geological history and the soil conditions of a location. This should be confirmed by available in-house data which might be exploration seismic data, top hole well log data, geotechnical data or existing survey data.

3.2 Timing

Adequate time must be allowed for planning and execution of rig site surveys. This will allow the survey data to be acquired, processed and interpreted with suitable quality control at all stages. Unfortunately rig site surveys may often have to be undertaken rapidly in response to short term changes in drilling campaigns. Due to the serious hazardous implications of sub-standard survey data and inadequate or improper use of results, insufficient time should not be allowed to compromise the quality of the survey results.

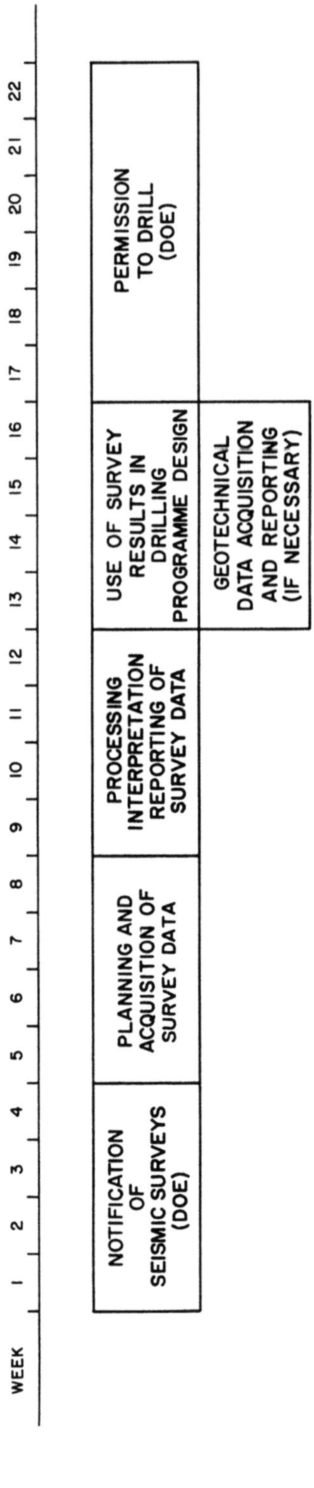

BAR CHART ILLUSTRATING RIG SITE SURVEY TIMING

FIGURE 1

Several factors should be considered when deciding upon the suitable timing for a rig site survey. These are illustrated by Figure 1. For either semi-submersible or jack-up rig site surveys, a minimum of 22 weeks should be allowed from initial survey planning to spud date. The timing of 22 weeks includes submission time to the Department of Energy, both pre and post survey, and the time to acquire, process and interpret the survey data. The timing given is considered realistic but to some extent will depend upon the individual operator's practice. If work has to be undertaken in less time, the quality of the survey results must not be jeopardized and the requirements of the Department of Energy must still be met.

3.3 Geophysical and Positioning Considerations

Many factors must be considered when deciding upon survey equipment and parameters, some of which are interdependent. Careful thought will ensure the right tools and procedures are selected for the job. These factors are discussed in detail in the procedures, particularly in the supporting appendices.

3.4 Additional Considerations

When planning site surveys, all other relevant data sources should be examined such as available geological, seismic, bathymetric and metocean data. Effective use of existing data will assist in targeting survey requirements.

3.5 Quality Assurance and Quality Control

Rig site survey data make a significant contribution to the safety of drilling operations. The quality of the data and interpretation cannot be taken for granted and require a great deal of careful quality control by specialists.

3.6 **Safety**

All injuries at work can be prevented and the operator should actively promote the high standards of safety consciousness, discipline and individual accountability that this principle demands.

When planning site surveys, safety must be given a high profile, wherever activity may take place.

3.7 **BGS**

Provision should be made for the site survey data to be sent to BGS upon completion, to be incorporated in the national geological archive.

4. SHALLOW GAS SURVEYS

4.1 **Objectives**

The objective of a shallow gas survey is to identify and map possible gas accumulations down to the sub-seabed depth at which normal well-killing procedures can be applied by means of a blowout preventer (BOP). This depth generally corresponds to the depth at which the 20 inch casing is set. This may be as great as 1000 metres sub-seabed, but will vary according to local geology and the individual operator's drilling practices.

Figure 2 has been included to illustrate the shallow gas problem. It is an example of seismic data from the Central North Sea acquired for

SHALLOW GAS EXAMPLE

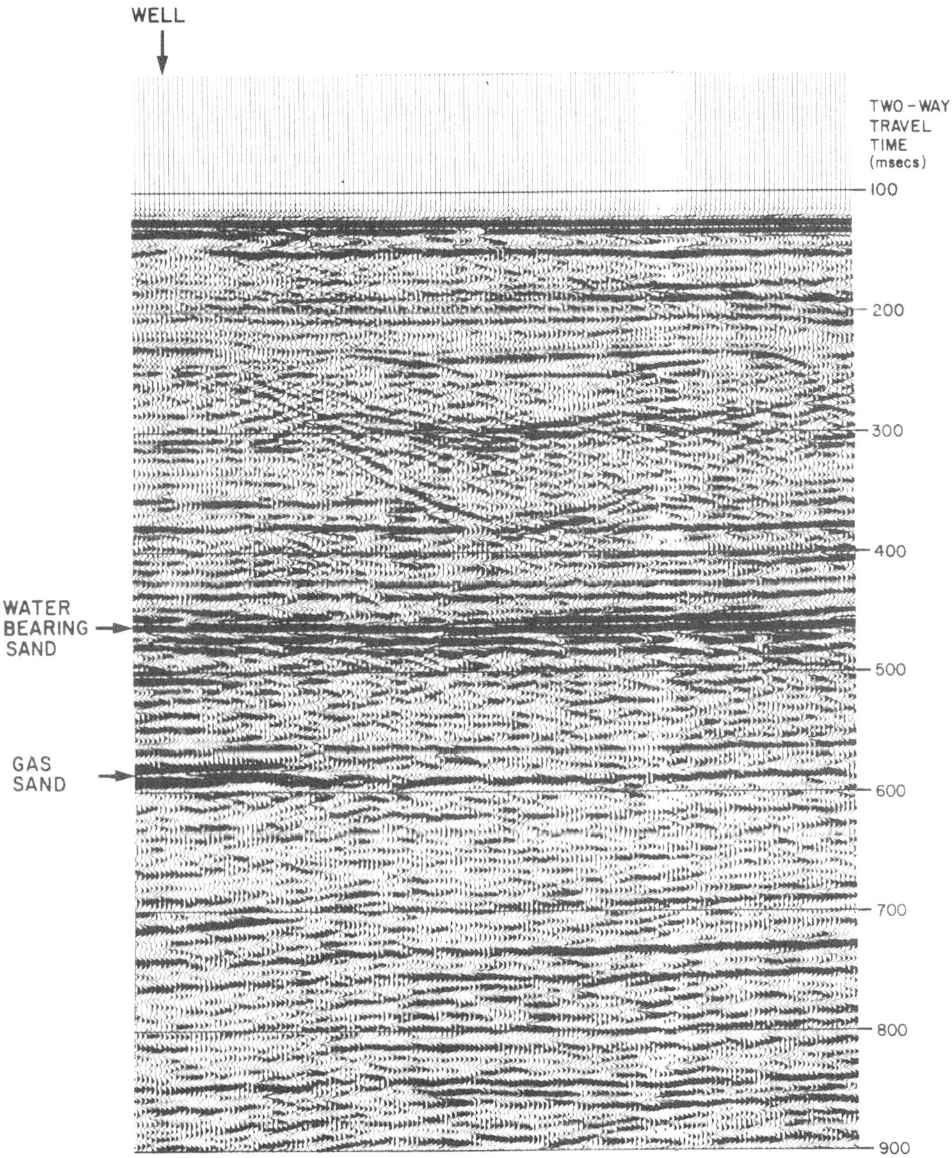

SLEEVE AIRGUN
48 FOLD
6.25m CMP
1ms SAMPLING

FIGURE 2

shallow gas detection. It shows an area of high amplitude reflectors at 470 milliseconds two way travel time and a very high amplitude reflector at 590 milliseconds two way travel time. The anomalous amplitude levels were drilled by an exploration rig and the upper level was proven to be a water bearing sand whereas the lower level was a gas charged sand. No drilling problems were encountered as the drillers were forewarned by the seismic survey and the gas pressures were easily controlled by normal drilling mud weight.

It is not possible to definitively set a lower limit to the thickness or areal extent of a gas charged unit which represents a drilling hazard. The objective of the survey is, therefore, to identify all gas accumulations within the 1000 metres sub-seabed interval. Practically, however, there are limits to vertical and lateral resolution, so a more realistic objective would be to detect (but not resolve the thickness of) potential gas accumulations of at least 1 metre thick, covering an area of 200 metres diameter at 1000 metres sub-seabed. Satisfying this objective will provide data capable of detecting smaller gas accumulations in the zone above 1000 metres sub-seabed.

Acquisition and processing of data should therefore be specified to provide optimum (technically and practically) vertical and lateral resolution throughout the objective interval.

Most gas accumulations should be detected. However, even with data acquired to the limits of available technology the results do not guarantee that shallow gas will not be encountered. Due to limitations in resolution and data interpretation, the technique should be viewed as a preventive measure, reducing the risk of encountering a gas accumulation. In addition, areas which are prognosed to be free of gas may later become gas charged as drilling may cause gas to migrate from elsewhere.

4.2 Equipment

To achieve the objective, multi-channel, digitally recorded, very resolute seismic data is required.

The parameters listed below provide minimum acquisition standards for shallow gas detection throughout the UKCS. Although local geological conditions may allow relaxation of the standards below, they should provide acceptable seismic data in most areas of the UKCS.

To illustrate the sensitivity of seismic data quality to acquisition parameters, Figure 3 shows a comparison of three data sets acquired for shallow gas detection in the Central North Sea. Although they are from the same area (within 2 kilometres of each other), the data sets are not directly comparable as they are not from seismic lines shot in precisely the same place.

The seismic data have been acquired with three different seismic sources and with different acquisition parameters and seismic processing. However, they illustrate very well the variability in results. The data to the right of the figure is acquired to the standards set out below. It provides a 'robust' data set with good signal-to-noise ratio throughout, which requires minimal seismic

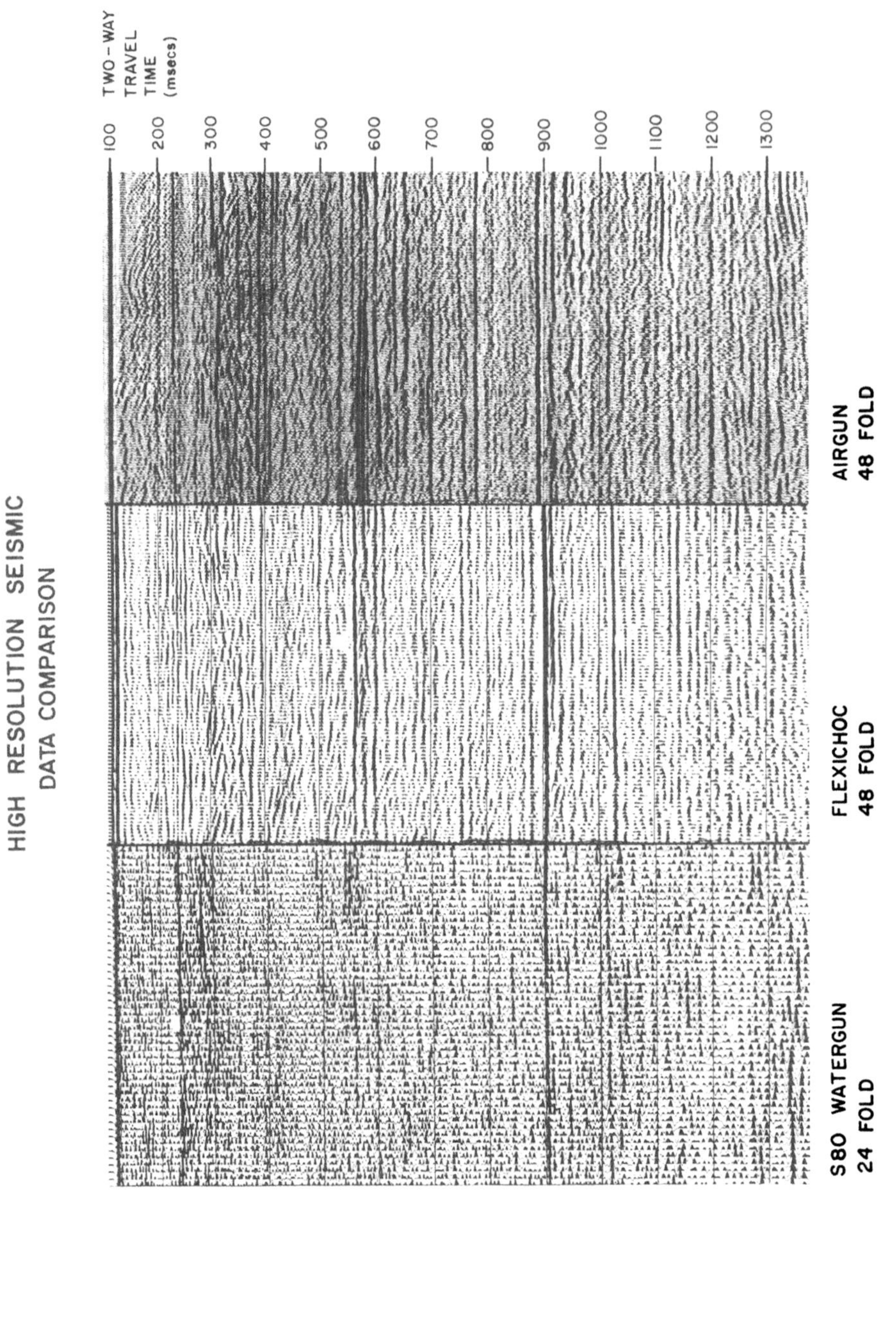
HIGH RESOLUTION SEISMIC
DATA COMPARISON
TWO-WAY
TRAVEL
TIME
(msecs)
100
200
300
400
500
600
700
800
900
1000
1100
1200
1300
S80 WATERGUN
24 FOLD
12.5 m CMP
0.5 ms SAMPLING
FLEXICHOC
48 FOLD
6.25 m CMP
1ms SAMPLING
AIRGUN
48 FOLD
6.25 m CMP
1ms SAMPLING

FIGURE 3

processing to achieve a satisfactory result. By comparison, the other two data sets show similar, or even a little better vertical resolution, but poor signal-to-noise ratio. In particular, the data set to the left of the figure is very noisy due to the seismic source characteristics, reduced multiplicity, and other acquisition parameters.

4.2.1 Seismic Source

The outgoing seismic signal should be stable, and at a tow depth of less than 3 metres, have the following minimum far field characteristics within the bandwidth of the operational filters:-

i) peak-to-peak amplitude in excess of 3.5 bar metres;

ii) bandwidth of at least 20 to 170 Hz at -6dB;

iii) primary to bubble ratio in excess of 15:1 (explosive sources);

iv) repetition rate of at least 3 seconds;

v) preferably minimum phase.

4.2.2 Hydrophone Streamer

i) minimum 48 groups;

ii) group length 6.25 to 12.5 metres;

iii) 600 metres minimum active length;

iv) sensitivity in the order of 5 volts per bar (transformer coupled);

v) depth controllers and indicators at least every 100 metres;

vi) preferably equal spacing of hydrophone elements throughout the streamer.

4.2.3 Acquisition Parameters

The recommended seismic acquisition standards for shallow gas surveys are tabulated in Table 1 below.

	MINIMUM REQUIREMENT	PREFERRED REQUIREMENT
Sample rate	1 millisecs	0.5 millisecs
Record length	1.5 secs	1.5 secs
Firing interval	6.25 m	6.25 m
Number of channels	48	96
Streamer group length	12.5 m	6.25 m
Source and streamer tow depth	2–2.5 m	1 m
High cut filter	≈ 360 Hz	≈ 720 Hz
Low cut filter	≈ 10 Hz	≈ 10 Hz
Source-near trace offset	minimum	
Streamer noise levels	5 microbars RMS maximum (this may be relaxed near depth controllers and at the front of the streamer)	
Feathering	6° maximum	

Table 1. Recommended seismic acquisition standards for shallow gas surveys

The preferred requirement of 0.5 millisecond sampled data requires a seismic source which will provide suitable high frequencies when towed at a shallow depth. Acquisition may be very difficult, due to the low weather tolerance of the shallow source and streamer tow depths. However, due to the improved vertical resolution, at least two lines over the drilling location should be considered using these parameters, in addition to the full grid of 1 millisecond sampled lines.

Figure 4 shows seismic data acquired to the minimum proposed, 1 millisecond sampled standard. This should be compared with Figure 5 which shows the same seismic line acquired to the preferred

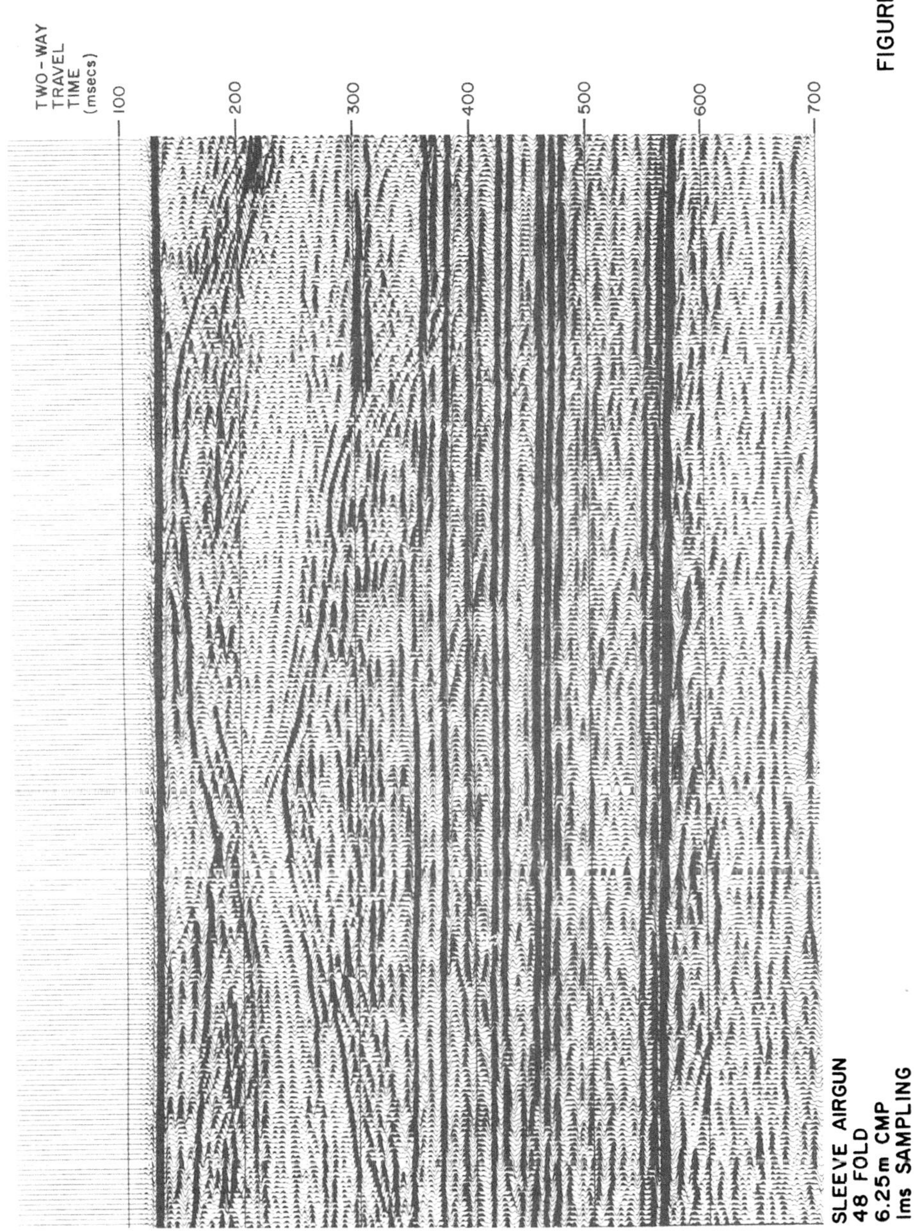
TWO-WAY
TRAVEL
TIME
(msecs)
100
200
300
400
500
600
700
SLEEVE AIRGUN
48 FOLD
6.25m CMP
1ms SAMPLING

FIGURE 4

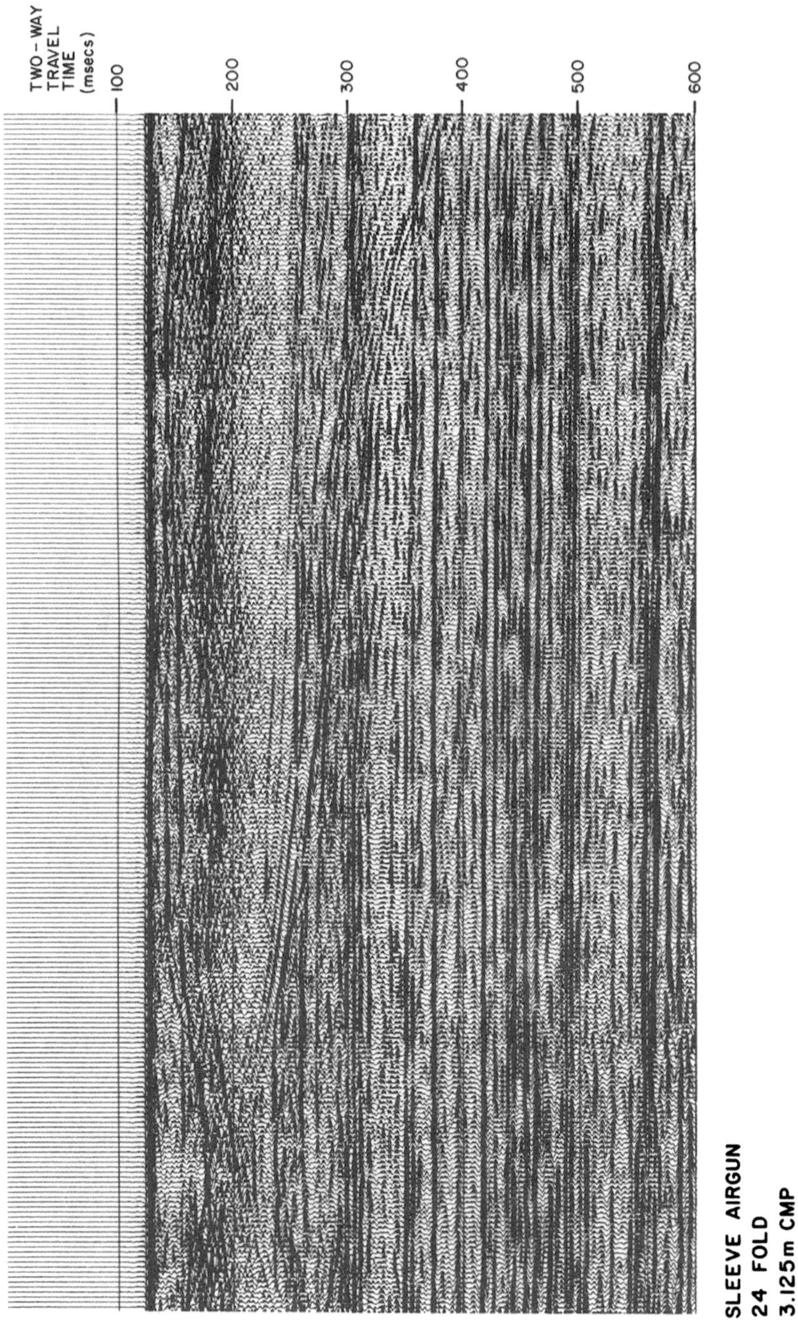
TWO-WAY TRAVEL TIME (msecs)
100
200
300
400
500
600
SLEEVE AIRGUN
24 FOLD
3.125m CMP
0.5ms SAMPLING

FIGURE 5

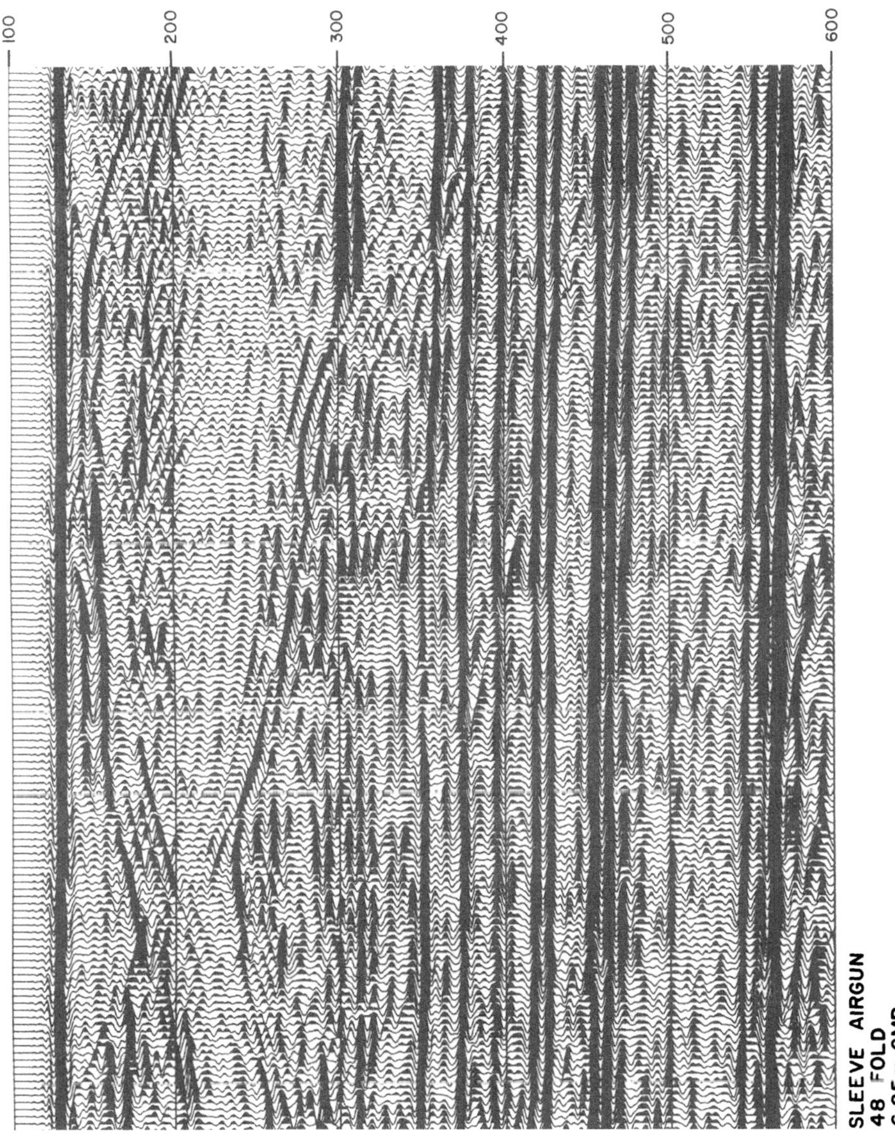
100
200
300
400
500
600
SLEEVE AIRGUN
48 FOLD
6.25m CMP
lms SAMPLING

FIGURE 6

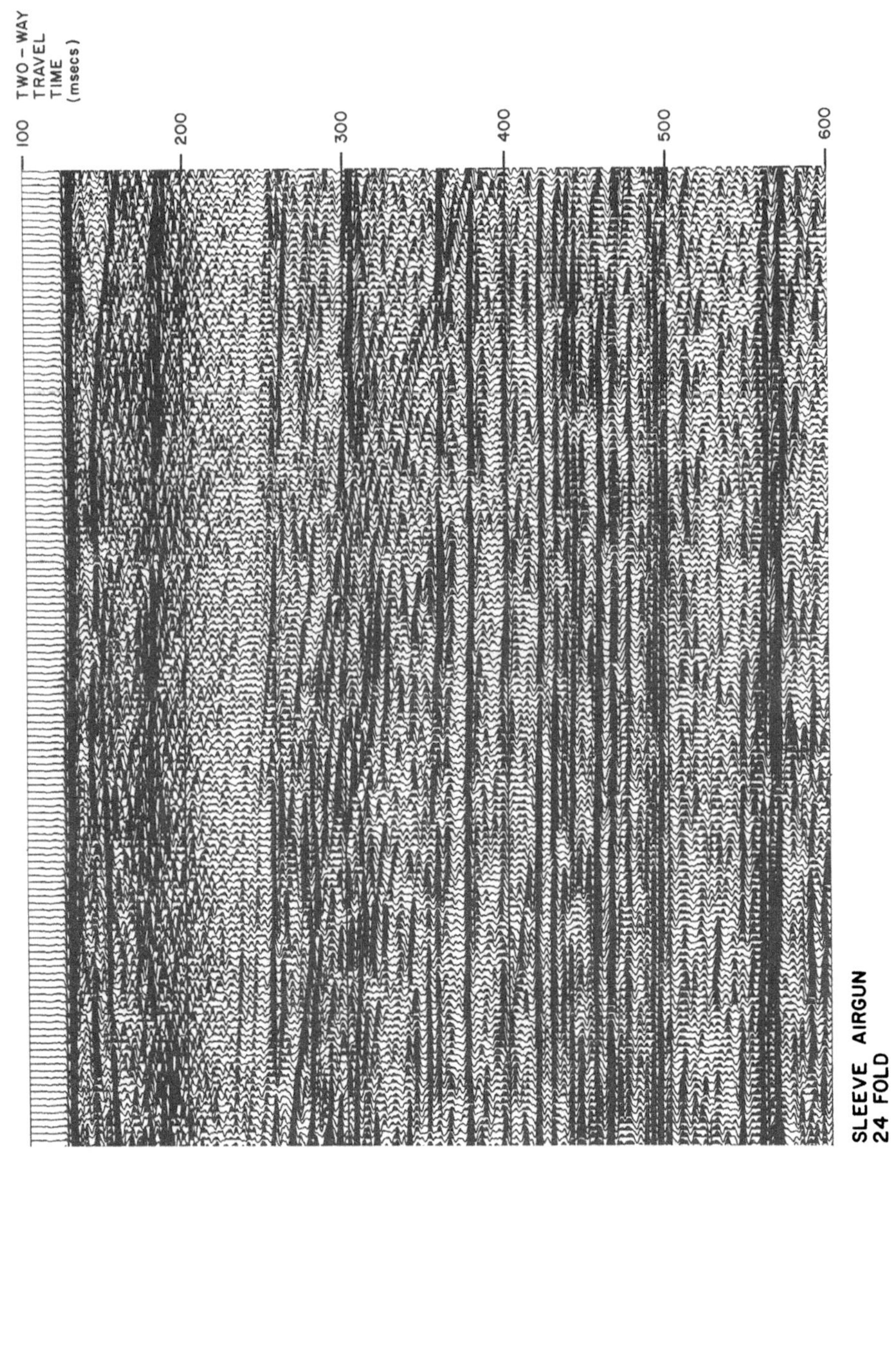
TWO-WAY
TRAVEL
TIME
(msecs)
100
200
300
400
500
600
SLEEVE AIRGUN
24 FOLD
3.125m CMP
0.5ms SAMPLING

FIGURE 7

0.5 millisecond sampled standard. The improvement in resolution, both vertical and horizontal, using the preferred standard is obvious and the ability to interpret anomalous events is enhanced. Figures 6 and 7 are sections of the same line at expanded scale to further illustrate the difference in resolution. Even at a two way travel time of 570 milliseconds, improved resolution is apparent.

4.3 Survey Pattern

4.3.1 Survey Area

Several factors must be considered when defining the survey area. If the final well location is not precisely known, the area surveyed should cover all possible locations. Should possible gas accumulations be detected, the survey area should be large enough that alternative locations may be identified. If the operator requires possible relief well locations to be surveyed, in case of blowout, a larger area will be required. If several closely spaced wells are likely, it may be more economical to perform one large survey, rather than several small ones.

Even if there is little flexibility to move the final well location, surveys of very small areas are not cost effective or useful. Very short seismic lines (a few hundred metres) are difficult to interpret and take almost as long to acquire as lines of reasonable length (in excess of 1 kilometre).

If top hole well logs are available from nearby wells, lines should be acquired to tie these wells to the location for correlation purposes. Even if no top hole well data is available, tie lines may be useful to relate drilling conditions at the proposed location to those at an already drilled location.

The recommended minimum survey area for shallow gas detection is 1 kilometre by 1 kilometre of full fold coverage, centred on the proposed location. This surveyed area should permit a recommendation for a move of location if necessary. However, features may be larger than 1 kilometre, so to assist in interpretation of the data, a few longer (several kilometres) lines should be considered. These could be tie lines to adjacent wells and/or lines through the proposed location.

4.3.2 Line Spacing and Orientation

Line spacing will dictate the ability to detect features of limited lateral extent. To ensure detection of small gas pockets at shallow depths, a line spacing of 50 metres around location is advised. This could be relaxed to 100 metres line spacing away from location. Should the location need to be moved away from shallow gas, to an area of 100 metres spaced lines, additional infill lines to 50 metres may be required, depending on local geology. Should no suitable drilling location be found within the 1 kilometre by 1 kilometre survey area, an additional survey would be required, but the longer lines may act as a basis for selection of the new location.

The survey grid should be orientated such that one seismic line heading is perpendicular to the geological strike and any fault planes. If the orientation of the lines has been selected to coincide with the heading of 2D exploration seismic data, depending upon the size of the drilling target, it may be possible to move the well location to an alternative position on the exploration seismic data without the requirement for deviated drilling. If 3D exploration seismic data is available, coincident line heading will not be required.

4.4 Data Processing

4.4.1 Seismic Data

The objectives of processing seismic data include improvement of signal-to-noise ratio, improvement of vertical and lateral resolution, suppression of multiple events, and enhanced display of data, hopefully in a quantified form.

Processing of data should be undertaken with extreme care since it is possible to create and destroy events on the seismic record. As a rule of thumb, the minimum processing of data should be undertaken and processing which destroys relative amplitude information must be avoided. Careful analysis of processing tests permits selection of those processing steps which enhance the data. When undertaking processing tests, the area selected should be representative of the site. Care should be taken to only alter one variable at a time so that the effect can be assessed. Processes should not be applied unless there is obvious benefit. On occasion, even though a process may be of obvious benefit in one way (e.g. multiple suppression) it may be disadvantageous in another (e.g. data suppression and noise creation).

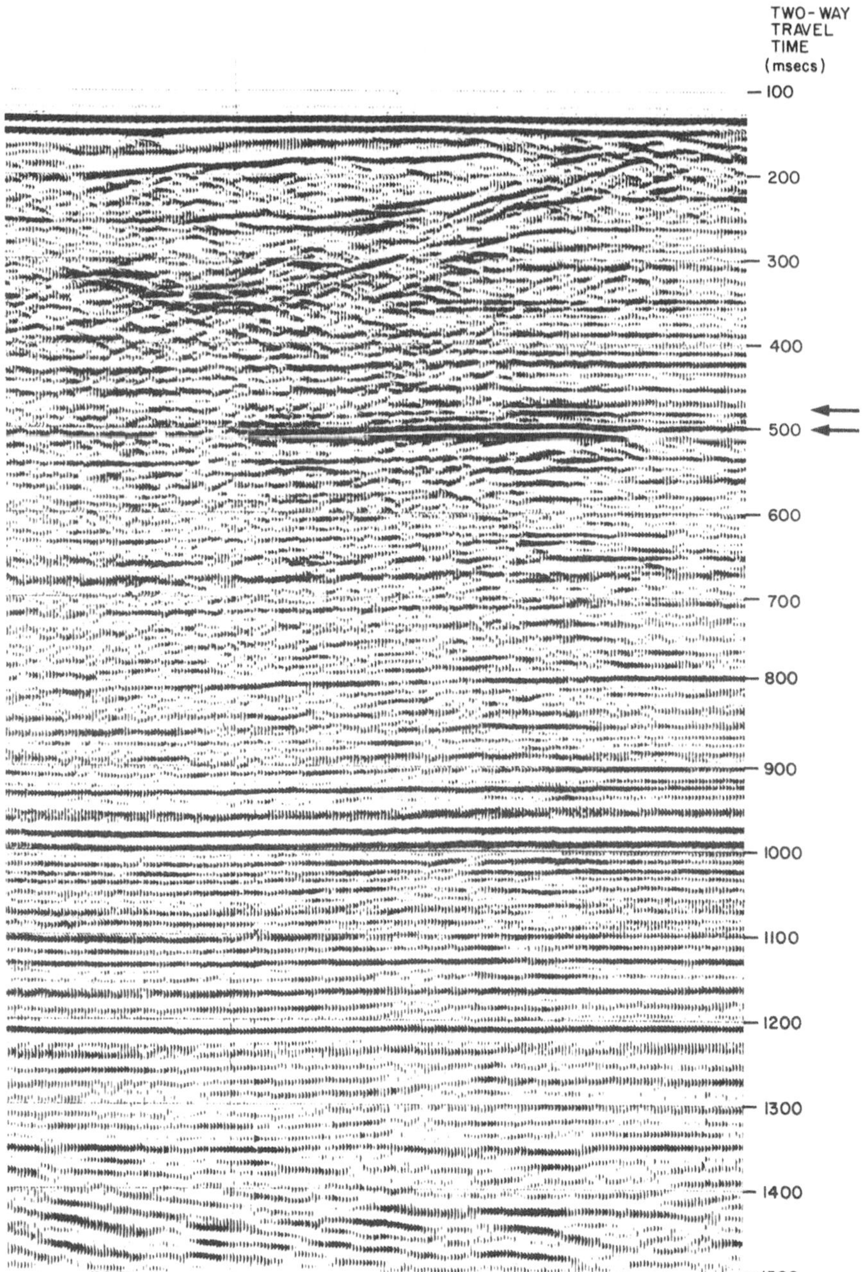

13 KJ SPARKER
24 FOLD
6.25m CMP
1ms SAMPLING

FIGURE 8

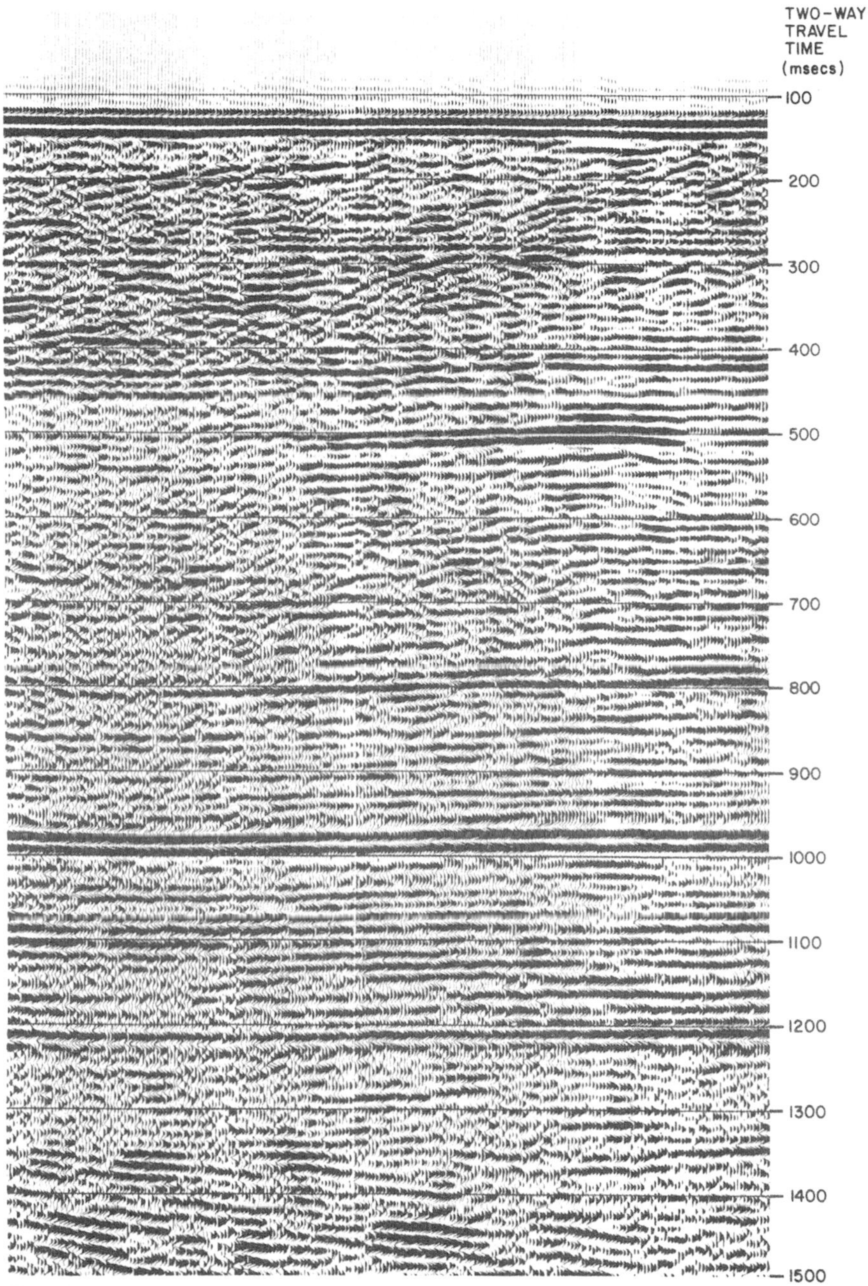

13KJ SPARKER
24 FOLD
6.25m CMP
1ms SAMPLING

FIGURE 9

Careful velocity analysis is of particular importance to ensure correct common mid-point stacking of the data. Velocity analyses should be undertaken at locations chosen with reference to the local geology. The minimum interval should be every 500 metres along the seismic lines. In some areas, where the geology is rapidly changing, (e.g. channels, salt domes), velocity analyses may be required more frequently than every 500 metres.

To illustrate the sensitivity of seismic data quality to processing, Figures 8 and 9 show a comparison of a shallow gas data set processed in different ways. The seismic data were acquired in the Central North Sea using a source which was not minimum phase. As the source was not minimum phase, careful wavelet processing was required to produce the results shown in Figure 8. Figure 9 shows the same seismic line processed, supposedly with care, by a reputable seismic processing company with the aim of providing data for shallow gas detection. Incorrect wavelet processing has seriously degraded data quality and, combined with incorrect filtering of the data, produced poor resolution. In particular the two levels of high amplitude reflectors, clearly seen on Figure 8 at 480 milliseconds and 500 milliseconds two way travel time, (see arrows), can only be clearly recognised as a single event on Figure 9. Also the channel flank and internal reflectors are poorly resolved on Figure 9. Incorrect seismic processing will seriously degrade data quality.

Portable seismic processing systems are now available, allowing seismic data processing onboard the vessel, sometimes on-line. These systems are very useful for quality control purposes. However, the onboard processed data should be used with extreme care for interpretation and shallow gas risk prognosis as the processing may be undertaken by personnel inexperienced in seismic processing, without adequate quality control, using processing and plotting packages which may be less sophisticated than those onshore.

4.4.2 Positioning Data

Positioning data requires processing on-line to allow the survey vessel to run along the required line pattern and for control of seismic shot interval and position. Raw positioning data should also be recorded to allow post-processing. Post-processing should remove spurious data, correct errors (if any) and take advantage of smoothing techniques. It also provides a quality control check on the real-time acquisition and provides information on potential systematic positioning biases.

Raw Positioning Data exchange should conform to UKOOA P2/86 format (soon to be revised), and Post Plot Positioning Data to UKOOA P1/84 format (soon to be P1/90).

Care should be taken to ensure all computed points (antenna, sensor) are based upon the correct offset.

4.5 Interpretation

Interpretation of data for shallow gas detection is subjective and should only be undertaken by experienced personnel. All sources of available data should be incorporated. Wherever possible this should include top hole well data, borehole data, exploration seismic data, adjacent site surveys, regional geological data (e.g. BGS and other literature), and in-house knowledge of the area.

4.5.1 Conventional Seismic Analysis

One characteristic seismic response of a gas charged sediment is a high amplitude reflection. Under certain conditions, low amplitude reflections are also possible. Theoretically the reflection should be phase reversed where gas is present. Other indicators of gas in sediments are masking and 'pull down' of underlying reflectors caused by the seismic wave passing through the lower velocity gas pocket. In practice these last two indicators are rarely seen as they are very sensitive to the geometry of the acquisition spread, with respect to the size and depth of the gas pocket. Consequently, identification of possible gas pockets depends primarily on identifying anomalously high amplitude reflectors.

High amplitude reflectors are caused by a strong impedance contrast. They may, therefore, be purely of lithological origin rather than due to gas accumulation. Constructive interference of reflected seismic waves from closely spaced reflectors may also cause high amplitude events (tuning effects), so the resolution of the seismic data is very important. The more resolute the data, the easier it is to discriminate genuine high amplitude events from tuning effects.

A small proportion (about 5 percent) of gas in a sediment will cause a high amplitude reflection of similar magnitude to that caused by a large proportion of gas (say 75 percent), so it is difficult to directly quantify gas content from the seismic response. Also, predictions of gas pressures based upon seismic data should be used with extreme caution.

4.5.2 Seismic Attribute Analysis

Seismic attribute analysis, including variation of amplitude versus offset effects (AVO), may assist in the discrimination of seismic events caused by shallow gas accumulations from those caused by lithology, e.g. lignite.

AVO effects are generated by changes in plane-wave reflection coefficients as a function of angle of incidence. In general, over the range of angles of incidence typical of shallow gas survey data, the top of a gas charged sand layer will show a marked increase of seismic amplitude with offset, whereas a water charged sand will show a small decrease of seismic amplitude with offset. Amplitude versus offset variations can be caused by many factors (e.g. reflection coefficient, array attenuation, tuning, noise, spherical spreading, absorption, emergence angle, reflector curvature, hydrophone sensitivity, and instrumentation and processing). However, with care, AVO analysis may assist in distinguishing gas related amplitude anomalies from other types of amplitude anomalies.

Increased reflector amplitudes, phase reversal, high frequency loss, and low velocities are seismic attributes which can be indicative of shallow gas accumulations and the ability to detect variations in them can be enhanced by attribute analysis and display. Variations in these attributes can also be caused by lithological variations and seismic processing, so interpretation based upon them should be undertaken with caution.

4.5.3 Calibration

Direct calibration of the data in the form of nearby top hole well data or borehole data is the most useful tool for assessing the significance of the seismic response and should always be carefully integrated with the data. In areas where no direct calibration is available, drilling departments should be encouraged to acquire top hole petrophysical data for correlation with the geophysical data and therefore improve confidence in the interpretation of an area.

Where the geological structure and interpreted lithologies provide a potential trap and no direct calibration is available, anomalous amplitude reflectors should always be assumed to be related to gas accumulations.

4.5.4 Depth Conversion

With care, the conversion from seismic two way travel time to depth below seabed, based on stacking velocities or interval velocities is accurate to better than 5 percent of the depth. Depth predictions can be further improved if top hole well data is available.

4.6 Reporting and Charting

Reporting should be concise and relevant to the survey objective. Consideration should be given to all potential users of the report and any information they require must be readily accessible. Inferences, unless substantiated by published data (e.g. BGS publications, results of previous wells) should be avoided.

Topics to be addressed by the report should include:-

i) Summary of Interpretation

ii) Interpretation

iii) Operations

iv) Calibrations

v) Data Reduction and Processing

vi) Equipment

vii) Data Quality

The objective of a shallow gas survey is to identify and map possible gas accumulations. The report must therefore clearly indicate the depth and lateral extent of any such accumulation. Any additional information to decide if the feature is a potential gas accumulation such as direct hydrocarbon indicators, structure, lithology should be included. Whenever possible, top hole well data should be used to substantiate data interpretation.

Charting should include maps showing the lateral extent of potential gas accumulations and interpreted sections. In addition, if there is a structure and potential gas accumulation this should be illustrated with charts showing depth to the interval below a horizontal datum (e.g. mean sealevel).

4.7 **Data Retention**

Data acquired during these surveys will include the following:-

i) Seismic Field Tapes

ii) Processed seismic data (paper, film and tape)

iii) Raw and Post Plot Positioning Tapes

All data should be retained at least until the well has been drilled and the drilling rig has moved to a new location.

It is recommended that all the data should be retained for a minimum of ten years after completion of the field work. In the event of further drilling activity in the area or field development, data can then be re-processed or re-interpreted if required using up-to-date or more refined techniques.

The British Geological Survey have requested that any rig site survey data (reports, maps, records and magnetic tapes) be sent to them, upon completion of the survey, to be incorporated in the national geological archive which they hold.

5. CONCLUSIONS

The proof version of the UKOOA Recommended Procedures for Mobile Drilling Rig Site Surveys has been reviewed by UKOOA members and other interested parties. Relevant comments are currently being incorporated.

The document provides procedures for a standard approach to rig site surveys with emphasis on good oilfield practice.

Recommended standards are proposed to ensure acquisition, processing and interpretation of seismic data suitable for shallow gas detection throughout the UKCS.

The UKOOA procedures also provide background documentation in support of the recommended standards which permits an understanding of the way in which results may be compromised if data is not acquired to these standards.

The quality of the data and interpretation cannot be taken for granted and requires a great deal of careful quality control by specialists.

Shallow gas surveys should be viewed as a preventive measure, minimising the possibility of encountering shallow gas.

ACKNOWLEDGEMENTS

The author wishes to thank Shell U.K. Exploration and Production and Esso Exploration and Production for permission to publish this paper. The author is also indebted to all those colleagues within Shell, UKOOA members and other interested parties who have reviewed the procedures.

REFERENCES

UKOOA Surveying and Positioning Committee, Seabed Survey Technical Sub-Committee (February 1990). Recommended UKCS Procedures for Mobile Drilling Rig Site Surveys (Geophysical and Hydrographic). (Draft).

Safety in Offshore Drilling -The Role of Shallow Gas Surveys

S.U.T. Conference held in London 25-26 April 1989.

Rapporteurs Summary By C D Green Shell Int Petr.MIJ

With acknowledgement and thanks to Dr A Dobinson B.G.S.

The conference was planned to coincide with the finalization and issue of the UKOOA mobile drilling rig site survey procedures.It was also intended to review the state of the art in shallow gas surveys and point up desirable and likely future trends in shallow gas surveys.
The conference was planned around opening introductory remarks from Dr H. Hughes of UKOOA> and a scene setting contribution from P Prince of BP Drilling Department.
Dr Hughes highlighted the importance of operators conducting their business safely and pointed out the essential requirement for an accurate prognosis of shallow gas conditions. A UKOOA study of the U.K. sector has indicated that some 1000 exploration wells and 2300 appraisal wells will be drilled in the next 25 years and that the development and refinement of a shallow gas prognosis capability is important and will continue to be a vital ingredient in offshore drilling.
P. Prince of BP Drilling stated that one third of all blowouts are caused by shallow gas. The prevention of shallow kicks and blowouts must rely initially upon a quality shallow gas survey and report. Relocation, directional drilling, careful selection and setting of casing depths, faster mud circulation and correct tripping techniques were all proposed as practical methods to reduce the chance of a shallow gas blowout.The role of riserless drilling was discussed together with a plea to ensure that gas surveys are planned and completed well in advance of spud date to fully assist in the formulation of the drilling programme.
Dr A. Judd questioned the importance of gas seepages and pockmarks pointing out that , in the medium to longer term life of an oilfield, wells and piles may act as a focal point to gas migration and escape.He suggested that operators should bear in mind the possible requirement for repeat surveys of gas prone areas to ensure that no migration and potential problems arise during the working life of an offshore installation.
R Price amply summarised the requirement for correct and proper source specifications pointing oput that just providing criteria for source level and p/b ratio was insuffient. He highlighted the importance of providing a specification including definition of the pulse in the frequency domain as a necessity to accurate source specification.

P. Newman (formerly of Horizon) discussed the application of the short offset 3D technique for high resolution site surveys. He questioned the need to shoot even low fold data as detail is lost in the spatial averaging of the stack process. Time slices from a small 3D survey were shown from the Ekofisk area revealing detail that would never have been possible to interpret from a 2D data set. The concept of low fold or even single fold data was proposed to optimize costs whilst producing a far superior dataset.Newman foresees a situation where single fold data would provide sufficient S/N and penetration relying on velocities generated from migration testing and application to permit depth conversion.
S. Middleton(Hydrosearch Associates) outlined the capabilities of onboard processing hardware detailing Micromax and Rameses In particular.it was suggested that the role of offshore processing should be in a Q.C. capacity,

Volume 25: Safety in Offshore Drilling, 291–293.

aimed solely at maintaining data integrity. The author agrees with this philosophy.
K. Games (Gardline Surveys) discussed the importance of the basic but most important routine steps of processing high resolution data. The paper conentrated on the disastrous results that can occur by incorrect velocity analysis and offset errors at the processing stage.
T. Armstrong(Total) discussed the findings of the use of AVO processing on a number of datasets.The case histories underlined the need for improved and careful pre-stack processing and it was proposed that the number of traces at short offset needs to be increased to allow effective AVO analysis of the shallow part of the section.
R. Salisbury (Britsurveys) presented an interesting case history paper showing shallow gas accumulations associated with salt induced tectonics. The association of bright spots with fault plane migration was clearly seen and the importance of accurate data migration was shown to map the gas transport paths.
B. Meier (Shell International) gave a detailed expose of the importance of adopting an integrated approach to shallow gas interpretation. The necessity to interpret the data in a geological framework involving use of exploration data was discussed together with the advice to use interactive workstation interpretation as standard to improve the quality of the shallow gas "prognosis" and thus accuracy.The use of individual and multi-attribute analysis should be regarded as an essential ingredient in the interpretation of shallow seismic data.
J. Williams (Hydrosearch Associates) described the results and benefits of the Signature project which was initiated in 1988.The objective of this was to produce proven diagnostic procedures for shallow gas detection in a work station environment.The summary of post stack diagnostics at any given target horizon include amplitude,instantaneous phase and frequency while below the target horizon blanking,pull down and drop in instantaneous frequency are shown.
R. Kanestrom (Norske Hydro) presented a paper on the prediction of shallow gas from seismic data. The paper concluded that the prediction of shallow gas is intimately linked to the parameters that are influenced most by changes in the porefill. The most significant factors that were cited include velocity,absorption,and amplitude versus offset. The absorption of seismic energy can be shown by plotting constant offset against true amplitude. AVO studies in themselves require more development to attempt to derive saturation parameters under ideal conditions.
A. Hill (BP) reviewing likely future developments in high resolution concluded that acquisition in the 2D sense was approaching a plateau and that perhaps 3D was the next and perhaps logical step forward.
In data analysis the paper suggested that oil companies must avoid a keyhole approach and provide appropriate feedback to the contractors to increase overall levels of expertise and experience within the industry.
P. Conn (Conn,Arthur and Associates) gave a comprehensive review of current practices and suggested that three future areas of special interest could be paying attention to instantaneous frequency changes that may be related to changing volume of gas, more development work on the use of AVO and a better understanding of pulse shaping variations caused by gas presence.
P. Walker (Shell UK repesenting the UKOOA survey committee) presented the outline of the forthcoming UKOOA procedures for mobile drilling rig site surveys. The objectives of these procedures are to provide a standardised approach to good oilfield practice. The procedures outline the minimum standards for data acquisition, identify pitfalls associated with data processing and comments on the methodology and limitations of interpretation. The procedures provide a comprehensive background to the

successful and safe completion of mobile drilling rig site surveysand whilst not intended as a text or reference book can be recommended to interested parties as extremely valuable and in formative reading!! Those concerned with the preparation of the procedures are to be congratulated.

The ensuing panel discussion consisting of Messrs Thomas (Britsurvey) Sommerville (Geoteam) and Darling (Gardline) were all invited by D. Ardus (Chairman) to comment on the proposed UKOOA procedures.All welcomed the proposed procedures commenting that they would welcome a unified approach and set of specifications. They further commented that it would also be helpful if a common standard and specification could be adopted for reporting. The question of liability once the procedures became the industry standard.was raised, but it must be recognized that the UKOOA procedures are for guidance purposes and the responsibility of individual operators to conduct their business according to good oilfield practice remains..

Concluding the conference as rapporteur C.D. Green (Shell International) did not attempt to summarize individual presentations,but took the opportunity to discuss the future after pointing out how much had been achieved in the site survey industry in the few years since the last technical conference held under the auspices of S.U.T. 2D acquisition was approaching a plateau and minds are turning towards high resolution 3D surveys. The quality of interpretation is generally improving bit it is hoped that by the time of the next SUT offshore site investigation conference in 2 years time work station interpretation and the use of attribute analysis would be the norm. As ever timing remains a problem and only by continuing pressure can individual operators ensure that site surveys are more adequately scheduled and completed

There remains much too be achieved in processing and interpretation techniques and in their application. The industry must develop ,adapt and adopt technology and techniques that can be derived from conventional exploration in order to improve the quality of , and quantify the shallow gas prognosis.

Final discussion concentrated on the possibilities and techniques of 3D high resolution surveys with great contrasts of opinion on the need for lower or single fold acquisition!

Overall the conference proved to be useful and informative with good attendance and a high level of participation by delegates. While the site survey industry cannot relax its much needed drive and desire to move forward technically it can take comfort in the knowledge that much has been achieved in the past few years to bring its credibility and technical base up to an acceptable standard.

C.D. Green Shell Int.Petr. MIJ. The Hague 16th May 1990.

GPSR Compliance

The European Union's (EU) General Product Safety Regulation (GPSR) is a set of rules that requires consumer products to be safe and our obligations to ensure this.

If you have any concerns about our products, you can contact us on ProductSafety@springernature.com

In case Publisher is established outside the EU, the EU authorized representative is:

Springer Nature Customer Service Center GmbH
Europaplatz 3
69115 Heidelberg, Germany

Batch number: 10607330

Printed by Printforce, the Netherlands